LEAN AND TECHNOLOGY

LEAN AND TECHNOLOGY

Working Hand in Hand to Enable and Energize Your Global Supply Chain

Paul A. Myerson

Associate Editor: Kim Boedigheimer
Senior Marketing Manager: Stephane Nakib
Cover Designer: Chuti Prasertsith
Managing Editor: Sandra Schroder
Senior Project Editor: Lori Lyons
Production Manager: Dhayanidhi
Copy Editor: Catherine Wilson
Proofreader: Sudhakaran
Indexer: Tim Wright
Compositor: codeMantra

For information about buying this title in bulk quantities, or for special sales opportunities (which may include electronic versions; custom cover designs; and content particular to your business, training goals, marketing focus, or branding interests), please contact our corporate sales department at corpsales@pearsoned.com or (800) 382-3419.

For government sales inquiries, please contact governmentsales@pearsoned.com.

For questions about sales outside the U.S., please contact intlcs@pearson.com.

1 16

ISBN-10: 0-13-429145-X
ISBN-13: 978-0-13-429145-1

Pearson Education LTD.
Pearson Education Australia PTY, Limited.
Pearson Education Singapore, Pte. Ltd.
Pearson Education Asia, Ltd.
Pearson Education Canada, Ltd.
Pearson Educación de Mexico, S.A. de C.V.
Pearson Education—Japan
Pearson Education Malaysia, Pte. Ltd.

Library of Congress Control Number: 2016952030

"Globalization is a fact, because of technology, because of an integrated global supply chain, because of changes in transportation."

—Remarks by President Barack Obama,
Joint Press Conference with Mexican President Enrique Pena Nieto,
White House East Room,
July 22, 2016

Contents at a Glance

Contents

Part III   Source 123

Chapter 9   Material Requirements Planning (MRP) 125

Chapter 10   Procurement (and e-Procurement) Systems 139

About the Author

Paul A. Myerson is a Professor of Practice in Supply Chain Management at Lehigh University and holds a B.S. in Business Logistics and an M.B.A. in Physical Distribution.

Professor Myerson has an extensive background as a Supply Chain and Logistics professional, consultant, and teacher. Prior to joining the faculty at Lehigh, Professor Myerson was a successful change catalyst for a variety of clients and organizations of all sizes, having more than 30 years of experience in Supply Chain and Logistics strategies, systems, and operations that have resulted in bottom-line improvements for companies such as General Electric, Unilever, and Church and Dwight (Arm & Hammer).

Professor Myerson created and has marketed a Supply Chain Planning software tool for Windows to a variety of companies worldwide.

He is the author of the books *Lean Supply Chain & Logistics* (McGraw-Hill, 2012) and *Lean Wholesale and Retail* (McGraw-Hill, 2014), Supply Chain and Logistics Management Made Easy (Pearson 2015), as well as a Lean Supply Chain and Logistics Management simulation training game and training package (Enna.com, copyright 2012–13).

Professor Myerson also writes a column on Lean Supply Chain for *Inbound Logistics Magazine* and a blog for *Industry Week* magazine.

PART I

Introduction and Overview

1

Lean Supply Chain and Technology: A Perfect Combination

Traditionally, Lean has been thought of as a "pen and pencil" technique for identifying and eliminating waste in business processes. As a consequence, while there are many books written on the topic of applying Lean tools and techniques in manufacturing, administration, and, to a lesser degree, supply chain, most if not all barely discuss the role and impact of technology in process improvement.

This book makes the case that technology is in fact a key enabler of a Lean supply chain and links Lean thinking with available and affordable systems and technologies to get the most out of improved processes.

Specifically, it details various supply chain, logistics, and operations management areas where Lean thinking (in its broadest possible sense) used in combination with existing and emerging systems and technologies—such as the Internet, e-commerce, enterprise resource planning (ERP) systems, 3D printers, bar code scanners, radio frequency identification (RFID) tags, etc.—can take an organization to the next level with increased speed, accuracy, integration, and collaboration among all parties in the extended, global supply chain.

We will discuss various tools, methodologies, best practices, examples, and cases of how, when and where technology can be combined with a Lean philosophy to "turbo charge" an organization's supply chain for a distinct competitive advantage.

It's a Small World After All

We live in exciting times, with the convergence of many activities and advances, including a global marketplace and supply chain, the growth of the Internet and e-commerce, omni-channel marketing and distribution, enterprise, and point software solutions coming in a variety of "shapes and sizes," and new hardware technologies for the gathering, analysis, and dissemination of information. This new world brings with it many risks and challenges as a result of the increased complexity from globalization, higher transportation costs, deteriorating or insufficient infrastructure, weather disasters, and terrorist threats. An organization's ability to navigate all this can give it a distinct competitive advantage; if these things are not managed well, it can mean potential failure.

All this helps to explain the increased interest in both process improvement methodologies such as Lean, Six Sigma, and the combined Lean Six Sigma and the increased use of technology to help enable and manage everything. However, as companies have limited resources, it is critical that they perform due diligence both in terms of process improvement and the selection and implementation of technology.

Like many other aspects of Lean thinking in the supply chain, technology can be an enabler of an improved process and can also help to retain and acquire new customers. So it's no surprise that technology, when used to collaborate with supply chain partners, can significantly reduce risk. This is explained quite clearly in the white paper "Mitigate Supply Chain Risk with Collaboration and Visibility to Achieve the Perfect Order," by Ariba, a procurement software vendor. This white paper states,

> In today's global economy, no company works alone. Intricate supply networks require interaction with hundreds—or even thousands—of outsourced resources, partners, suppliers and customers around the world. That's why innovative manufacturers are embracing new collaboration and automation technologies to help overcome inefficient, error-prone, manual processes. By enabling collaboration without boundaries, industry leaders are providing greater visibility into direct materials to deliver the perfect order. [www.ariba. com, 2014]

The white paper points cites a recent survey of supply chain executives, which found that mitigating risk is the single biggest day-to-day challenge and one of the highest priorities when collaborating with trading partners. The survey also showed that almost all respondents found value in collaboration and that it's a top priority for most chief procurement officers. The collaboration can take many forms, including sharing information, automating processes, and using business networks or hubs. Most of those surveyed felt that manual processes were a major concern (and source of waste), and the majority of those surveyed also recognize that a new generation of technology is critical for automating processes, enhancing collaboration, and mitigating supply chain risk.

Ultimately, those who have used technology for visibility and collaboration attain significant benefits, such as improved order accuracy and faster, more automated transaction cycles—all signs of a leaner, more efficient supply chain.

How Big an Opportunity Is It?

A variety of studies (and my own experience) shows that anywhere from 50% to 95% of Lean programs fail. While a huge reason is the lack of an ingrained Lean culture, it can also be that the technology a company has, doesn't have, or eventually selects isn't a good match for its current or future processes. This can result in wasted time, money, and effort.

Many studies have shown that Lean can potentially result in significant (over 50%) improvements in areas such as cycle time, inventory, capacity, and quality. While Lean thinking can certainly create these types of improvements, modern technology enables these improved processes, serving as a kind of "lubricant" for the entire supply chain—not only institutionalizing the improvements but in many cases taking them further than first thought possible.

When businesses think of spending money on technology, which can range from 1% to 7% of total revenue (with manufacturing and retail at the lower end and financial and health care services at the higher end), it's usually to benefit them in terms of the ability to:

- Reach more potential customers and better service new and existing customers
- Develop a tighter relationship with existing or potential customers, suppliers, and other key partners
- Improve or streamline operations, resulting in reduced costs and wastes, improved efficiency, and greater profits

At least partially as a result of these types of perceived benefits, spending on information processing equipment and software increased from 18.2% of all business investments in 1987 to a peak of 46.7% at the end of 2000 [Pisello, 2001]. This has continued to increase, and a TD Bank survey of chief financial officers (CFOs) in 2015 found technology to be 58% of total capital spending by companies in 2016, followed by existing facilities (44%) and data security (41%) [www.tdbank.com, 2015].

It should come as no surprise then that worldwide information technology (IT) spending—including includes hardware, software, IT services, and telecom markets— totaled $3.8 trillion in 2015. Of this total, $335 billion was spent on enterprise software, $732 billion on devices (including personal or mobile computers, mobile phones, tablets, and printers), $143 billion on data center systems (servers and storage and network equipment), $981 billion on IT services, and $1,638 billion on telecom services [Gartner, 2015].

The Extended and Often Global Supply Chain and Technology

As the number of partners and length of shipment times increase, so do the degree of complex, multi-enterprise interactions and the need for seamless integrated visibility and responsiveness across multiple enterprises. A lack of automation and visibility handcuffs companies with longer lead times, bigger inventory buffers, budget over-runs, and continued demand–supply imbalances. Three-quarters of respondents in an Aberdeen Global Benchmark survey reported that they don't have enterprise-wide automation for global supply chain processes. On average, large companies said that their global supply chains are only 50% as automated as their domestic supply chains. Furthermore, 79% of large companies said that the lack of supply chain process visibility was their top concern, while 90% of all enterprises reported that their global supply chain technology was inadequate to provide the corporate finance organiza-tion with the timely information it requires for budget and cash flow planning and management. To address this, companies are moving away from building in-house applications and moving toward using packaged applications [Aberdeen Group, 2006].

A more recent Aberdeen Group survey found that "best practice" organizations have taken control of their own supply chains as they tend to be situated in the middle of dozens of supply chain partners, where they share data and adopt universal standards for bar coding, track and trace, and data exchange. That control may generate more information but also requires more collaborative technology.

Gaining visibility at all levels of product and shipment details and to all trade and financial costs and transactions can expose problems and opportunities. However, when they enable adaptive and collaborative (Lean) processes, leading organizations have superior cost, service, and competitive advantage. Still, there is plenty of room for improvement, as 63% of respondents felt that supply chain visibility was a high priority for improvement [Aberdeen Group, 2013].

Information Systems (IS) Versus Information Technology (IT)

The terms *information systems* (*IS*) and *information technology* (*IT*) are often used interchangeably. Traditionally, IS refers to manual and/or computerized systems that are designed to create, modify, store, and distribute information and that consist of people, processes, machines, and information technology, while IT usually deals with the technology part of any information system, such as hardware, servers, operating systems, software, and so on. It's not surprising then that both IS and IT can help support and enable Lean thinking in an organization.

Risks and Rewards

While selecting and implementing systems and technology in an organization can be justified for a variety of reasons, such as improved productivity, inventory turns, and customer service levels, as discussed throughout this book, many risks need to be considered as well. For example, a survey on the success of ERP systems implementations by Panorama Consulting Systems [www.zdnet.com, 2013] showed that most ERP projects run over budget, and buyers do not fully receive expected benefits. Some respondents felt that the ERP projects were failures. Specifically, the survey showed that the following:

- More than 50% of projects experienced cost overruns.

- More than 60% of projects experienced schedule overruns.

- 60% of respondents received less than half of the expected benefit from their ERP implementation.

So both separately and together, Lean thinking with IS and IT can have a huge impact on an organization; they can also be risky propositions that suck up vast resources and yield less-than-anticipated returns. My belief is that technology can enable a good process. So it's only natural that a continuous improvement tool like Lean as part of a supply chain strategy can, and should, go hand in hand with technology—but only if done in the right way so that the investment of scarce resources (capital and human) are spent wisely. Therefore, one of the main purposes of this book is to help link the two together—to show the synergies between them and describe methodologies for determining which tools and technologies have the greatest impact in an organization.

Linking Competitive Strategy to the Value Chain

Historically, supply chain and operations management functions were viewed primarily as cost centers to be controlled. In recent years, it has become clear that these functions can be used to gain a competitive advantage that helps the top line as well. An organization should establish competitive priorities that its supply chain must have in order to satisfy internal and external customers. It should then link the selected competitive priorities to its supply chain and operations management processes.

Krajewski et al. suggest breaking an organization's competitive priorities into cost, quality, time, and flexibility capability groups (see Figure 1.1) [Krajewski, 2013]:

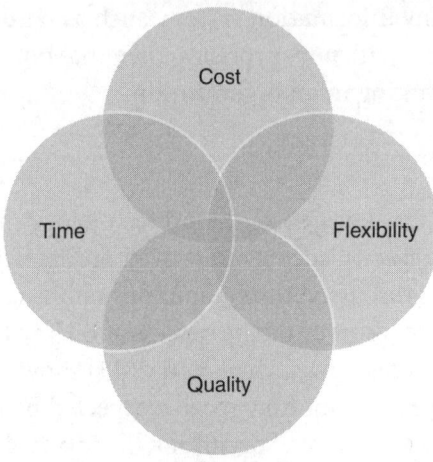

Figure 1.1 Competitive Priorities

- **Cost strategy**—This strategy focuses on delivering a product or service to the customer at the lowest possible cost without sacrificing quality. Walmart has been the low-cost leader in retail by operating an efficient supply chain.

- **Time strategy**—A time strategy can focus on speed of delivery, response time, and consistency or even product development time. Dell has been a prime example of a manufacturer that has excelled at response time by assembling, testing, and shipping computers in as little as a few days. FedEx is known for fast, on-time deliveries of small packages.

- **Quality strategy**—Consistent, high-quality goods or services require a reliable, safe supply chain to deliver on this promise. If Sony had an inferior supply chain with high damage levels, it wouldn't matter to the customer that the company's electronics are of the highest quality.

- **Flexibility strategy**—This strategy can focus on priorities such as volume, variety, or customization. Many of today's e-commerce businesses, such as Amazon, offer a great deal of flexibility in many of these categories.

Many organizations focus on more than one of these strategies, and even those that focus on only one of them must offer reasonable performance in the others (though perhaps not "best in class" performance).

The goal for today's supply chain is integration through collaboration to achieve visibility downstream, toward the customer, and upstream, to suppliers. In a way, many of today's companies have been able to "substitute information for inventory" to achieve efficiencies. The days of having "islands of automation" or having an internally focused internal system that may optimize one organization's supply chain at the cost of someone else's (such as a supplier's or customer's) are over.

The Lean philosophy involves teamwork and critical thinking aided by the right technology to enable organizations to work with other functions internally as well as with other members of the external supply chain, including customers, suppliers, and partners. The organization can then achieve new levels of efficiency and use its supply chain to achieve a competitive advantage by focusing on adding value to the customer as opposed to just being a cost center within the organization.

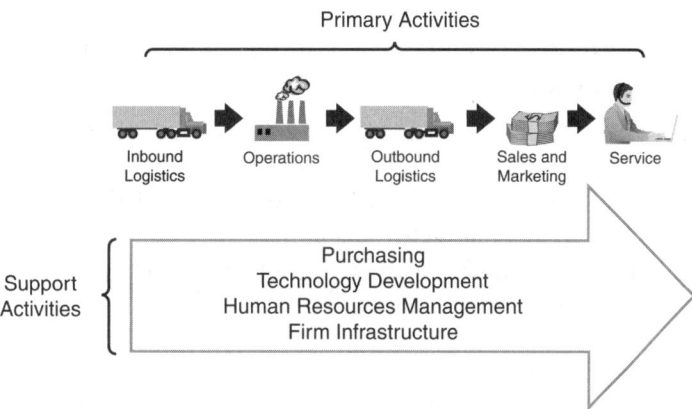

Figure 1.2 Value Chain Model

The Value Chain model, originated by Michael Porter (which today is more like a "Value Web," as each businesses' chain intersects with other chains), shows the value-creating activities of an organization. As you can see in Figure 1.2, it relies heavily on supply chain functions.

In a value chain, each of a firm's internal activities, listed below, adds incremental value to the final product or service by transforming inputs to outputs:

- **Inbound logistics**—Activities including receiving, warehousing, and inventory control of input materials

- **Operations**—Activities related to transforming inputs into the final product or service to create value

- **Outbound logistics**—Actions that get the final product to the customer, including warehousing and order fulfillment

- **Marketing and sales**—Activities related to buyers purchasing the product, including advertising, pricing, distribution channel selection, and the like

- **Service**—Activities that maintain and improve a product's value, including customer support, repair, warranty service, and the like

Porter also identifies support activities that can add value to an organization:

- **Procurement**—Purchasing raw materials and other inputs that are used in value-creating activities

- **Technology development**—Research and development, process automation, and similar activities that support value chain activities

- **Human resource management**—Recruiting, training, development, and compensation of employees

- **Firm infrastructure**—Finance, legal, quality control, and so on

Porter recommended Value Chain analysis to investigate areas that represent potential strengths that can be used to achieve a competitive advantage. As shown in the Figure 1.3 example of a manufacturer, the supply chain adds value in a variety of ways, so it should be a critical area of focus.

The Value Chain model also includes linkages between the activities. For example, sales forecasts drive production, and production determines raw material and component needs. The tighter the linkages between these activities, at least in theory, the

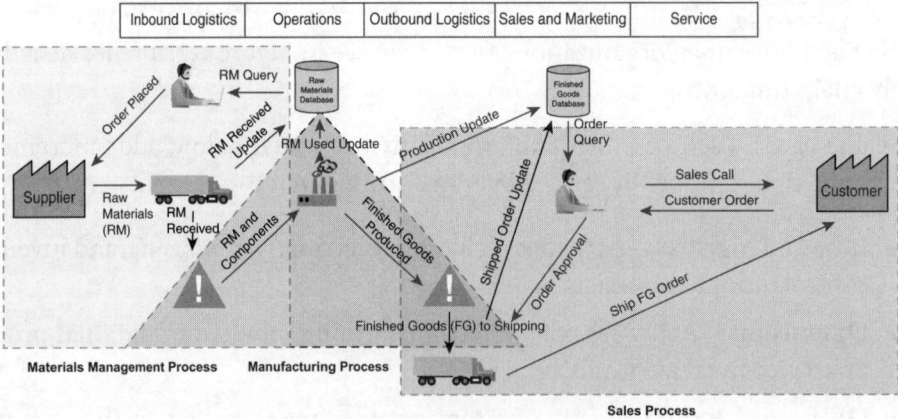

Figure 1.3 Manufacturer Processes, Linkages, and Information Flow

lower the inventory levels and associated costs. This model therefore was the driver for all encompassing business systems that crossed functions in an organization. It also led to business process redesign (BPR), which brings us back to the idea that just automating current (less-than-optimal) processes isn't necessarily the best idea. Rather, new and more integrated processes should be created with linkages between the value-added activities [Porter, 1985].

Competitive Strategy, Business Processes, and IT Structure Aligned

When an organization selects a competitive strategy, it designs business processes that include and link value-generating activities. The processes themselves will determine the organization's IT requirements. It is therefore critical that businesses align their IT with their business objectives. Doing so requires them to do the following:

- Identify business goals and objectives.
- Break strategic goals into value activities and processes.
- Identify specific measures to define success.
- Decide how IT can help to achieve business goals.
- Measure actual performance versus goals.

In Figure 1.3, the specific IT needed for the manufacturer would depend on its competitive strategy as well as where it feels it does or can add the most value.

For example, if the company is a low-cost business, it would want systems that help find and buy lower-cost raw materials and components, perhaps leveraging the Internet more on the procurement side to locate and negotiate with those type of suppliers. It might also want to utilize bar code technology to help speed orders and inventory through its finished goods warehouse. An example of this in the retail world is Walmart, which uses information systems to achieve the lowest operational costs and the lowest prices. Walmart's inventory replenishment system sends orders to suppliers when purchases are recorded at cash registers; this helps the company minimize inventory at warehouses, reduce operating costs, and collaborate with manufacturers to improve forecast and replenishment speed, cost, and accuracy.

On the other hand, a company that has a more flexible, responsive strategy, inventory turns might not be a priority, but inventory availability, shorter lead times, and customer service might be. As a result, such a company might want systems and technology that help perform quicker changeovers on manufacturing equipment, automated distribution management systems for orders to be processed more quickly, and tight integration with key customers to determine its ever-changing product and service requirements. Companies utilizing this type of competitive strategy, such as Dell Computer, may use information systems to customize and personalize products to fit specifications of individual consumers. At Dell's e-commerce site, customers select the options they desire and order computers custom built to their specifications in a short amount of time. Dell's assemble-to-order system gives the company a significant competitive advantage.

Other strategies include:

- **Focusing on a market niche**—An organization may use information systems to enable specific market focus and serve a narrow target market better than competitors. This strategy involves analyzing customer buying habits and preferences and making advertising pitches to smaller and smaller target markets.

- **Strengthening customer and supplier relationships**—An organization may establish strong links to customers and suppliers while increasing switching costs and loyalty. Automakers use IS to facilitate direct access from suppliers to production schedules so that they can improve their forecasts and schedules. Many e-commerce sites keep track of user preferences for purchases and suggest titles to existing customers.

As in many other aspects of the business world, with competitive strategy there are multiple ways to do things. For example, Dell strives for both low cost and a responsive, customized product, and it is very successful, with inventory turns of 90 times per year and a lead time for a customized computer of around 4 days.

In addition, the Internet has had a huge impact on competition and has transformed many industries. The Internet has reduced the costs of operating globally, enabling smaller businesses to compete on a global scale and helping to establish new products and services with faster time to market. It has also increased the bargaining power of customers and suppliers through transparency and scope and speed of communication.

Using IS to gain a competitive advantage requires coordination of people, process, and technology, which as will discuss throughout this book (see Figure 1.4).

According to Ramakrishman and Testani, "Far too often business transformation efforts concentrate on the process improvement strategies and business process reengineering; while essentially ignoring the people aspect of the change initiative.

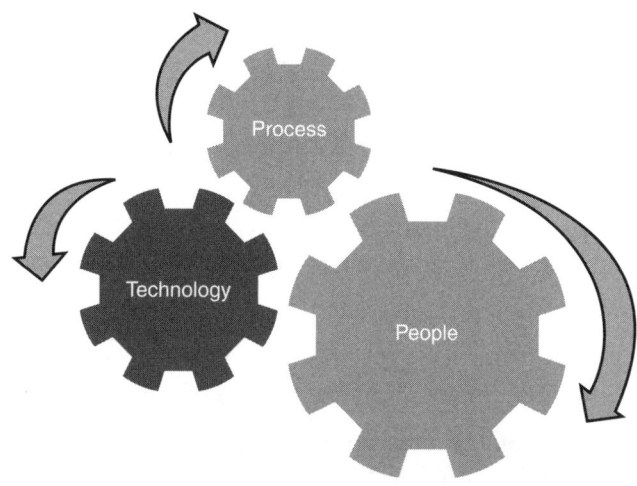

Figure 1.4 People, Process, and Technology

Subsequently, these transformation initiatives do not achieve their desired results. Studies have shown that approximately three quarters of business re-engineering efforts do not achieve their objectives and subsequently do not sustain themselves over the long term, and one of the most commonly cited reasons for their failing is due to the lack of focus on the organization's culture." By aligning people, process, and technology, companies can develop "critical organizational competencies around organizational culture transformation and process improvement; resulting in a more effective and sustainable change effort" [Ramakrishman and Testani, 2011].

Using the SCOR Model to Help Enable Lean Opportunities with Technology

We will be discussing the interaction of supply chain and operations processes with technology throughout this book, and the most organized way to do this, I believe, is using the SCOR model (see Figure 1.5). This model, which was designed by the Supply Chain Council (which has since merged with APICS; see www.apics.org), divides the supply chain into six management processes:

1. **Plan**—This process involves balancing supply and demand, which we will discuss in detail along with the sales and operations planning (S&OP) process. These plans are communicated throughout the supply chain.

---

2. **Source**—This process involves the procurement of goods to meet demand. It includes identification, selection, and performance measurement of sources of supply, as well as delivery and receipt of materials.

3. **Make**—This is the transformation process, which involves converting raw materials into finished products.

4. **Deliver**—This process involves the resources needed to move materials along the supply chain, from suppliers to manufacturing and then to customers. It includes order management, warehousing, and shipping.

5. **Return**—This is the reverse logistics process for product or material that is returned, including repair, maintenance, and overhaul.

6. **Enable (added in SCOR version 11 in 2012)**—This process focuses on supporting the other five process steps with best practices to support those steps.

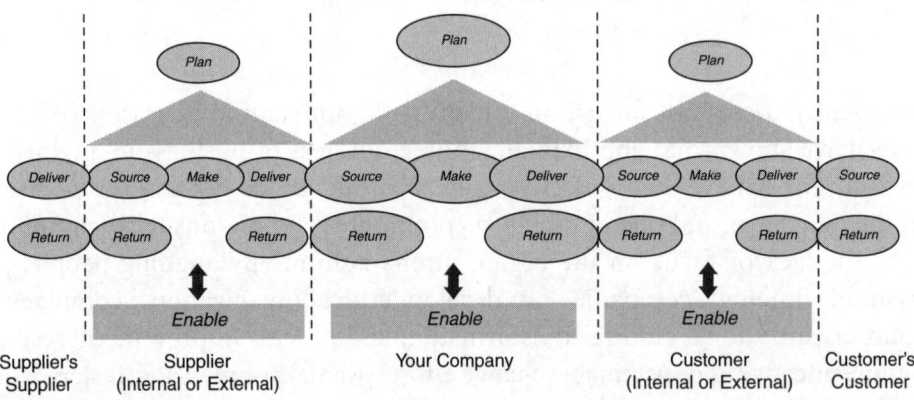

Figure 1.5 SCOR Model

Before we get into specifics about applying Lean tools and technology to various processes, it is important to gain some insight into how and why the supply chain has grown in importance, understand some Lean concepts and tools, and discuss the general technology sourcing process.

2

The Growing Importance of the Supply Chain and Technology to Business and Society

Supply Chain Defined

The terms *supply chain* and *supply chain management* (*SCM*) should be separately defined as they are sometimes (mistakenly) used interchangeably.

The *supply chain* is a system (actually more like a web than a chain) of organizations, people, activities, information, and resources involved in the planning, moving, or storage of a product or service from supplier to customer. Supply chain activities transform natural resources, raw materials, and components into a finished product that is delivered to the end customer.

Supply chain management, as defined by the Council of Supply Chain Management Professionals (CSCMP), "encompasses the planning and management of all activities involved in sourcing, procurement, conversion, and logistics management. ... It also includes the crucial components of coordination and collaboration with channel partners, which can be suppliers, intermediaries, third-party service providers, and customers." Supply chain management integrates supply and demand management within and across companies and typically "includes all of the logistics management activities noted above, as well as manufacturing operations, and it drives coordination of processes and activities with and across marketing, sales, product design, finance and information technology" [www.cscmp.org, 2014].

Some people take a narrower view of supply chain. Some think of it as being focused more on the supply end (i.e., purchasing) than the logistics side (i.e., the part of the supply chain that plans, implements, and controls the efficient movement and storage of goods, services, and information from the point of use or consumption to meet customer requirements). Others assume that logistics is included in the supply chain but don't state it. Still others, while including both areas above, ignore the planning aspects of the supply chain. Personally, I tend to refer to the field as *supply chain and logistics management* to make clear what is included.

In fact, according to the 2011 article "Continuing Education—Making the Right Selection" by Perry A. Trunick, "Some of the most passionate debates in academic circles still center on what constitutes supply chain management and its place in the academic structure. Not surprisingly, that same debate rages in the commercial world" [Trunick, 2011]. The article goes on to say that some people use the terms *logistics* and *supply chain* interchangeably, while others feel that it's important for logistics to still have its own place.

It is very important to understand the similarities and differences between more functional areas such as logistics, which includes transportation and distribution, and the broader concept of SCM, which is cross-functional and cross-organizational. This can have a major impact on decision making, structure, staffing, and technology requirements in an organization, so it needs to be understood and examined carefully.

Depending on one's view, some of the following functions may be included within the supply chain and logistics organization:

- **Procurement**—The acquisition of goods or services from an outside external source
- **Demand forecasting**—Estimates of the quantity of a product or service that customers will purchase
- **Customer service and order management**—Tasks associated with fulfilling an order for goods or services placed by a customer
- **Inventory**—Planning and management
- **Transportation**—For hire and private
- **Warehousing**—Public and private
- **Materials handling and packaging**—Movement, protection, storage, and control of materials and products using manual, semi-automated, and automated equipment
- **Facility network**—Location decisions in an organization's supply chain network
- **Information management (including generating and sharing customer, supplier, forecasting, inventory, and production information)**—All the information required to ensure that supply matches demand throughout the supply chain

Supply chain management is also intertwined with operations management, which consists of activities that create value by transforming inputs (i.e., raw materials) into outputs (i.e. goods and services). We therefore sometimes use the term *supply chain <u>and</u> operations management* (see Part IV, "Lean Supply Chain and Technology: Make.") as both activities support the manufacturing process.

History of Supply Chain and Logistics Management

What is now known as *supply chain management* has a long history, though the term wasn't coined until the early 1980s. It started with a focus on cost minimization for organizations and now encompasses interconnected, complex global networks, the efficiency of which can determine the success or failure of a company.

Frederick Taylor created scientific management operations research techniques that were used on military logistics processes in World War II. After World War II, businesses began to understand the relationships and trade-offs involved, such as inventory costs versus transportation costs, and logistics gained an important place in the business world as well.

In the 1960s, physical distribution, a more integrated concept which included activities such as transportation, inventory control, warehousing, and facility location, had become an area of study and practice in education and industry. Physical distribution involved the coordination of more than one activity associated with supplying product to the marketplace (i.e., more focus on the "outbound" side of manufacturing).

In the mid-1960s, the scope of physical distribution was expanded to include the supply side, including inbound transportation and warehousing, and was referred to as *business logistics*. In many cases, purchasing was not included but instead fell under materials management or procurement.

In the early 1980s, American manufacturing had been fighting overseas competitors for over a decade, and U.S. companies began actively outsourcing materials, labor, and manufacturing overseas. At this point, the term *supply chain management* entered the common business lexicon. It defined both the new, complex global world we now live and do business in and an understanding of the integration and importance of all activities involved in sourcing and procurement, conversion, and logistics management. Supply chain management includes coordination and collaboration with channel partners, which can be suppliers, intermediaries, third-party service providers, and customers.

In the past, physical distribution, logistics, purchasing, and so on were fragmented. Today, many companies have an integrated supply chain organization, in most cases led by a senior level executive (see Figure 2.1).

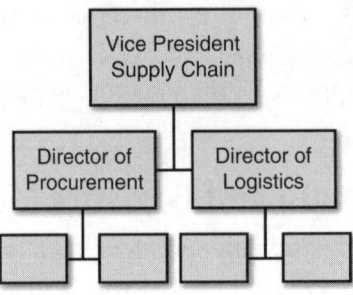

Figure 2.1 Supply Chain Organizational Chart

Technology has helped drive the concept of an integrated supply chain. For example, electronic data interchange (EDI) systems were developed in the 1980s as a standardized format for the electronic transfer of data between business enterprises. In addition "off-the-shelf" enterprise resource planning (ERP) software systems feature integrated core business processes in a common database. An important development in the 21st century has been the expansion of Internet-based collaborative systems.

Since the mid-1960s, U.S. companies have also been slicing up their supply chains in search of low-cost and capable suppliers offshore. In 1980s there was a move from "producer-driven" supply chains to "buyer-driven" chains. Today, global supply chains cover finished goods as well as components and sub-assemblies, in both goods and services industries. Today, more than 28% of U.S. GDP is tied to trade, with world exports of intermediate goods exceeding the combined export values of final and capital goods. Total annual U.S. logistics costs alone are estimated to be well over $1 trillion, with over 50% of that spent on transportation costs and 33% on inventory carrying costs. [U.S. Chamber of Commerce]

Intricate supply networks require interaction with hundreds—or even thousands—of outsourced resources, partners, suppliers, and customers around the world. Innovative manufacturers are therefore embracing new collaboration and automation technologies to help overcome inefficient, error-prone manual processes.

This supply chain evolution has resulted in both increasing value added and cost reductions through integration and collaboration, with a wide range of technologies enabling this to occur efficiently. Specifically, the benefits from supply chain technology can include the following:

- Added competitive advantage
- Increased visibility

- Increased efficiency
- Better customer relationships
- Optimized operations
- Improved responsiveness
- Better decision making
- Supply chain operations optimization

The Role of Information Technology in the Supply Chain

There really isn't any aspect of supply chain and logistics that isn't touched by technology in today's world. Today, thanks to both advances in software and hardware technologies and the Internet, companies of all sizes can automate and integrate internal processes and connect with customers and suppliers with ease.

Supply Chain Information

Information provides the foundation on which supply chain processes execute transactions and managers make decisions. Hardware, software, and people throughout a supply chain gather, analyze, and execute upon information. The information must be accurate, accessible in a timely manner, of the right kind, and shared. Information is used when making decisions about facilities, inventory, transportation, sourcing, and even pricing and revenue management (which are sales and marketing responsibilities that are impacted by supply chain structure and efficiency).

Interactive View of Information

There are subtle differences between data and information. *Data* are the facts from which information is derived to make decisions. Pieces of data are rarely useful alone; for data to become information, they need to be put into context. That is the purpose of information systems.

According to Simatupang and Sridharan [2001]:

> An interactive view of information enables people to define the level of information they need to solve problems or make decisions. Depending on the decisions, some people can use data to answer the questions, but others need to extract information from the same data to solve their problems. This

interactive view also enables people to trace the source of knowledge from the available data, or to specify the required data based on their explicit knowledge.

An information system is used to collect, process and disseminate information to make it available for decision makers at the right time. Traditionally, an information system deals with transferable data through plain media of communication such as EDI and the internet. The recent advance of information technology offers a rich variety of media such as video conferencing and online decision support systems that enable decision-makers to convert tacit knowledge into explicit knowledge and to share explicit knowledge.

Figure 2.2 shows the interactive view of information described by Simatupang and Sridharan.

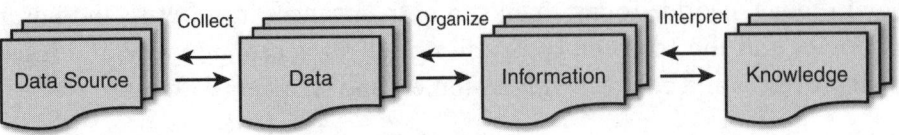

Figure 2.2 Interactive View of Information

In general, an organization needs information to be easy to access, relevant, accurate, and timely. Thus, the information technology used has a direct impact on a company's performance, both internal and external through integration, which enables collaboration (see Figure 2.3).

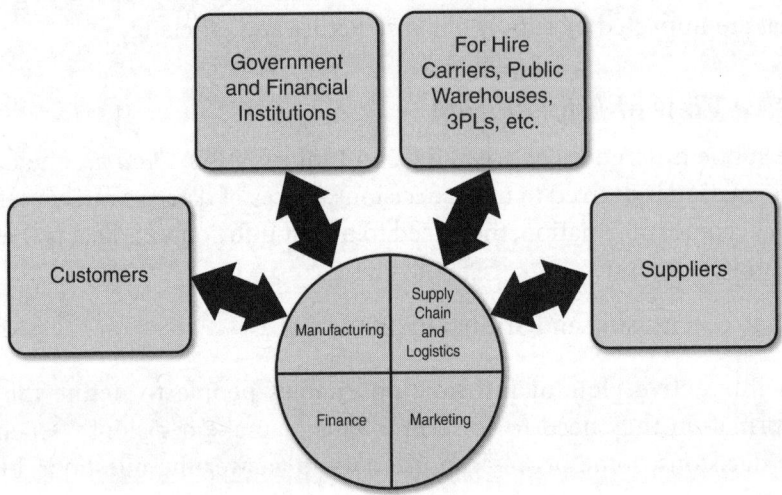

Figure 2.3 Internal and External Supply Chain Information Flows

Importance of Integration for Sharing Information

A phenomenon known as the bullwhip effect (see Figure 2.4) occurs when demand variability increases, moving up the supply chain away from the consumer, and small changes in consumer demand can result in large variations in orders placed upstream. It is largely the result of poor information management in the supply chain and therefore can lead to excess inventory levels. By having greater demand visibility throughout the supply chain, a company can reduce its inventory levels. Therefore, at least in theory, it is possible to substitute information for excess inventory through the use of information systems.

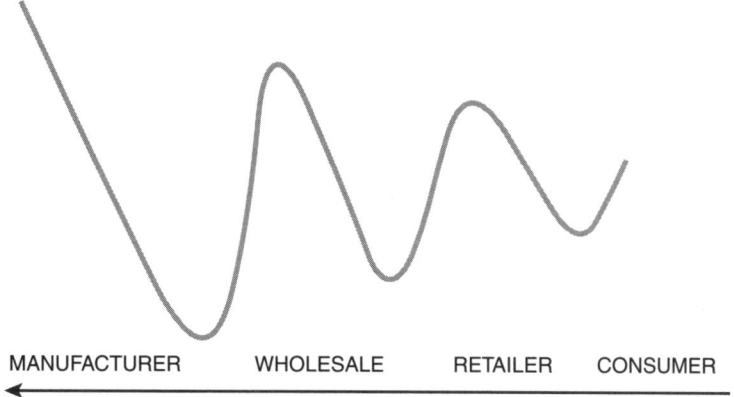

Figure 2.4 The Bullwhip Effect

In terms of the supply chain itself, Fawcett and Magnan (2002) identified four levels or steps of integration:

1. Internal cross-functional integration
2. Backward integration with first-tier suppliers
3. Forward integration with first-tier customers
4. Complete backward and forward integration (from the supplier's supplier to the customer's customer)

Moving from the traditional, fragmented supply chain to a more integrated internal and external supply chain requires changes to the ways both customer and supplier relationships are established and managed.

In order for a supply chain to achieve its maximum level of effectiveness and efficiency, material, money, and information flows throughout (internal and external) must be integrated and managed, within overall service and cost objectives. To accomplish this, information technology (IT) must play a major role.

Viewpoints of Supply Chain Information Systems

Since there are so many applications in today's global supply chain, it is best to look at information needs from strategic, tactical, routine, and execution viewpoints:

- **Strategic**—This viewpoint develops long-term decisions that help meet the organization's mission and focus on strategic plans for meeting them, such as new products or markets, as well a facility capacity decisions.

- **Tactical**—This viewpoint develops plans that coordinate the actions of key supply chain areas, customers, and suppliers across the tactical time horizon. These plans focus on tactical decisions, such as inventory or workforce levels. They plan but don't carry out the actual physical flows.

- **Routine**—This viewpoint supports rules-based decision making, usually in short time frames, where accuracy and timeliness are important to the user.

- **Execution**—This viewpoint is typically transaction oriented, recording and retrieving transaction processing data and executing control over physical and financial information flow. These systems usually have very short time frames, are highly automated, and use standardized business practices [Bozarth, 2008].

Supply Chain Macro Processes

One way to look at the supply chain and its functional technology needs, at least at a high level, is by breaking it into the following macro processes:

1. **Supplier management (SM)**—SM ensures that supplies are at the best cost and terms. SM can range from a strategic buy, to a tactical negotiated purchase, to a heavily engineered item.

2. **Internal supply chain management (ISCM)**—ISCM includes a number of activities related to receiving, conversion, and movement of finished goods.

3. **Distribution channel management (DCM)**—DCM involves the links in a distribution network that has multi-tier arrangements. It depends on the industry and type of products shipped and can also include service providers such as transportation, distribution, and third-party logistics (3PL) companies.

4. **Customer relationship management (CRM)**—CRM includes the practices, strategies, and technologies that companies use to manage and analyze customer interactions and data throughout the customer lifecycle, with the goal of improving business relationships with customers, assisting in customer retention, and driving sales growth.

5. **Transactional management (TM)**—TM involves the base transactional data, such as order and inventory information, to run the day-to-day aspects of a business.

As shown in Figure 2.5, processes 1–4 provide access and reporting of supply chain transaction data. Advanced systems use analytics based on transaction data to improve supply chain performance, and enterprise resource planning (ERP) systems form the foundation of a supply chain IT system.

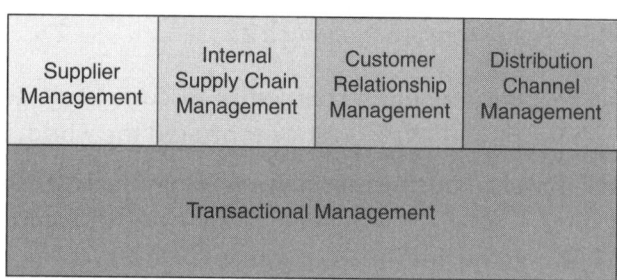

Figure 2.5 Supply Chain Macro Processes

Supply Chain Information Technologies

On a more practical level, supply chain management (SCM) systems can be also viewed in terms of planning (SCP) and execution (SCE) technologies tied to processes across the supply chain:

- **SCP**—Applies algorithms to predict future requirements of various kinds and to balance supply and demand and can include systems for demand management, supply management, and sales and operational planning (S&OP) to ensure that supply matches demand.

- **SCE**—Monitors physical movement and status of goods as well as the management of materials and financial information of all participants in the supply chain. SCE can include systems such as warehouse management systems (WMS), transportation management systems (TMS), and, of course, enterprise resource systems such as ERP.

There are also information technologies for the following:

- **Supply chain event management**—Systems used for managing events that occur within and between organizations or supply chain partners. The goal is to keep all users in the supply chain—from materials suppliers and buyers to warehouse managers and product carriers—informed about activity across the supply chain. These systems typically perform event monitoring, notification, simulation, control, and measurement processes.

- **Business intelligence (BI)/supply chain analytics**—Applications, infrastructure, tools, and best practices to help turn data into actionable information through analysis to improve and optimize decisions and performance. These systems can include reports, real-time dashboards, and benchmarking.

Information links all parts of the supply chain, and the technology involved includes a number of hardware and software tools such as:

- **Internet**—The Internet allows companies to communicate with suppliers, customers, shippers, and other businesses around the world, instantaneously.
- **E-business**—Physical business processes are gradually being replaced with electronic ones. There are two general forms of e-business: business-to-business (B2B) and business-to-consumer (B2C).
- **Electronic data interchange (EDI)**—EDI is the computer-to-computer exchange of standardized business documents. Today, EDI may also occur through the Internet.
- **Bar code and point-of-sale data**—Such data can be used to create an instantaneous computer record of a sale.
- **Radio frequency identification (RFID)**—RFID is technology that can send product data from an item containing an RFID chip to a reader via radio waves.

Supply Chain Technology Trends

Technology is having a profound effect on current supply chain processes, and it is evolving so rapidly that it may actually lead to process changes. A Capgemini supply chain management trend study found that technical innovation will have a tremendous impact on the supply chain of tomorrow [Capgemini, 2014]. It found that the following elements will have the greatest impact:

- **Emerging technologies**—Functionality is being developed in the area of global communication and information flows within the supply chains, particularly machine-to-machine communication supported by sensors and content information. Production processes will be accelerated, and more transparency will be available to support management and operations to make informed decisions.

- **Analytics and simulation**—In the future, supply chain decisions will be based on real-time information instead of assumptions.

- **Supply chain segmentation**—Instead of the "one fits all" mentality used in the past, different customers associated with different channels and different products are being served through different supply chain processes, policies, and operational modes. Ultimately it's about finding the best supply chain processes and policies to serve each customer and each product at a given point in time while also maximizing both customer service and company profitability. Simulation technology is used to define the ideal supply chain for each segment. Transactional technology is also used to identify relevant supply chain events and act on those interferences, to ensure agreed service levels specifically in high-speed segments.

- **Service orientation**—Service orientation is related to a segmentation strategy. The goal is to make sure that agreed service-level agreements are met. To be successful, global supply chain event information should be integrated centrally, enabling the supply chain control group or teams to react to supply chain events quickly and make sound decisions.

- **Optimization**—There are already a variety of supply chain network optimization software tools, and they continue to become faster, more accessible and easier to use. Optimization based on operations research methodologies is the most advanced and mature area in supply chain management IT-supported functions.

- **Sustainability**—Sustainability can have positive effects on the cost side. More efficient return and recycle processes, reductions in energy consumption, and waste-avoiding processes lead to significant cost reductions. Planning and optimization technology can help achieve sustainable and efficient network design to help supply chain planners define efficient, sustainable supply chain structures that reduce energy use and cost.

While the growth and importance of the global supply chain has at least partially been enabled by technology, it has created a much more complex environment in which to operate. The concept of Lean supply chain, covered in Chapter 3, "Lean Concepts and Their Applications in the Supply Chain," has gained momentum to help to effectively deal with this challenge.

3

Lean Concepts and Their Applications in the Supply Chain

Lean Background

Lean thinking has been around for a long time in one form or another. It stems from a management philosophy used by Japanese automobile manufacturer Toyota soon after World War II. That philosophy is known today as the Toyota Production System (TPS), and it is important because it shifted the focus of manufacturing from individual machines and their utilization to the flow of the product through the entire process.

TPS focused on having appropriate-sized machines for the volume required as well as machines that in essence were self-monitoring to ensure quality and process sequence. Toyota also originated the idea of quicker equipment setups to enable production of small volumes of a variety of items, as well as the concept today known as a *kanban*, where each downstream process lets upstream processes know of its need for materials. The TPS philosophy provided Toyota low-cost <u>and</u> high-variety production along with high-quality and quick throughput times, which meant the company could respond to changing customer demand—referred to today as *demand "pull" manufacturing*. Pull stands in stark contrast to the concept of "push" manufacturing, based on the concept of economies of scale, where large quantities of single items are produced to spread fixed costs over many units in order to keep the cost per unit low.

Since the 1990s, this process, used in many organizations outside the automotive industry, has been known as *Lean*. Womack and Jones [1996], who helped make the term *Lean* part of the popular lexicon, distilled its principles into the following:

- Specify the value desired by the customer.
- Identify the value stream for each product providing that value and challenge all the wasted steps (generally 9 out of 10) currently used to provide it.

- Make the product flow continuously through the remaining value-added steps.
- Introduce pull between all steps where continuous flow is possible.
- Manage toward perfection so that the number of steps and the amount of time and information needed to serve the customer continually fall.

Essentially, Lean is a team-based form of continuous improvement that identifies and eliminates non-value-added activities or waste through a relentless focus on exactly what the customer wants. While it was originally used in the automobile and other repetitive manufacturing industries, in recent years, the easy-to-understand and easy-to-implement concepts and tools of Lean have spread to other forms of manufacturing and into supply chain and logistics, services, retail, health care, construction, maintenance, and government.

Value-Added Versus Non-Value-Added Activities

In order to understand the Lean concept of waste, it is first important to understand the meaning of value-added versus non-value-added activities.

Any process entails a set of activities. The activities in total are known as *cycle time* or *lead time*. The lead time required for a product to move through a process from start to finish includes queues/waiting time and processing time.

The individual activities or work elements that actually transform inputs (e.g., raw materials) to outputs (e.g., finished goods) are known as *processing time*. In general, processing adds value from the customer's standpoint. Processing time is the time it takes an employee to go through all of his or her work elements before repeating them. It is measured from the beginning of a process step to the end of that process step.

Consider a simple example of making raw lumber into a pallet of 2 × 4s. In this case, the value added to the customer is the processing that transforms the raw lumber into the final pallet of 2 × 4s. This process includes activities such as washing, trimming, cutting, and so on, which are a relatively small part of the cycle time; that is, it may take only one hour to process the raw material into a finished pallet, but the entire cycle time may be one week.

In Lean terms, the non-value-added time is actually much greater than just the lead time. It includes current inventory "on the floor" (i.e., raw, work-in-process, and finished goods), and we can use a calculated takt time for a specific "value stream" (a single product/service or family of products/services, as discussed later in this chapter) to convert those quantities to days of supply. Doing so can expand the non-value-added time from days to weeks (or even months).

It is very common for many processes (or value streams) to have only 5% to 10% value-added activities (see Figure 3.1). However, there are some non-value-added necessary activities, such as regulatory, customer required, and legal requirements. These activities don't add value, but they are necessary, so we can't eliminate them, but we should try to apply them as efficiently as possible to avoid waste.

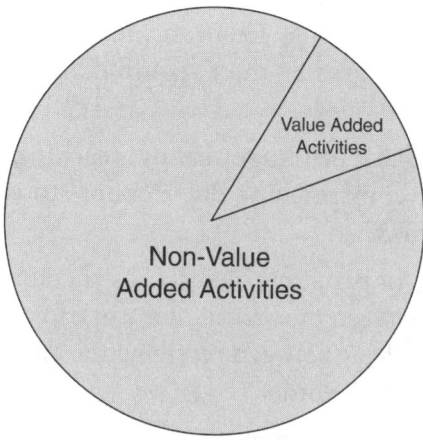

Figure 3.1 Value-Added Versus Non-Value-Added Activities

It is fairly normal for management to focus primarily on speeding up processes—often value-added ones, such as the stamping speed on a press. With Lean, the focus moves to speeding up non-value-added activities, and in some cases this may even result in slowing down the entire process in order to balance it, remove bottlenecks, and increase flow.

Waste

In Lean, non-value-added activities are referred to as *waste*. Typically, when a product or information is being stored, inspected, delayed, waiting in line, or defective, it is not adding value and is 100% waste. Such waste can be found in any process, whether it's manufacturing, administrative, supply chain and logistics, or elsewhere in an organization.

Eight wastes are typical, and one easy way to remember them is that they spell "Tim Woods" (see Figure 3.2):

- **Transportation**—Excessive movement of people, products, and information. This may include out-of-route stops, excessive backhaul, and locating fast-moving inventory in the back of a warehouse, which may cause unnecessary material handling distances.

- **Inventory**—Storing material or documentation ahead of requirements. Excess inventory often covers for variations in processes as a result of high scrap or rework levels, long setup times, late deliveries, process down-time, and quality problems. This can include early deliveries, receipt of order for a quantity greater than required, and inventory in the wrong warehouse.

- **Motion**—Unnecessary bending, turning, reaching, and lifting, which may include excess travel or reaching due to poor storage arrangement or poor design of work areas.

- **Waiting**—Waiting for parts, information, instructions, or equipment. This may include the time between the arrival of a truck for a pick-up and the loading of the trailer or the delay between receiving the customer's order information and beginning to pick the order.

- **Overproduction**—Making more than is immediately required. This causes excess inventory or paperwork, potentially resulting in an increase in any or all of the other wastes.

- **Overprocessing**—Tighter tolerances or higher-grade materials than are necessary. In the supply chain, for example, the longer the time required to process orders, the long the time to ship the product and get paid.

- **Defects**—Rework, scrap, and incorrect documentation (i.e., errors). This includes product quality issues, and also, in the supply chain, data errors can result in shipping orders late, incorrectly, and at increased expense. It can also result in higher inventory costs to try to compensate for these inaccuracies.

- **Skills**—Underutilizing capabilities of employees and delegating tasks with inadequate training. This is perhaps the greatest waste of all, as employees are a company's greatest asset and source of ideas.

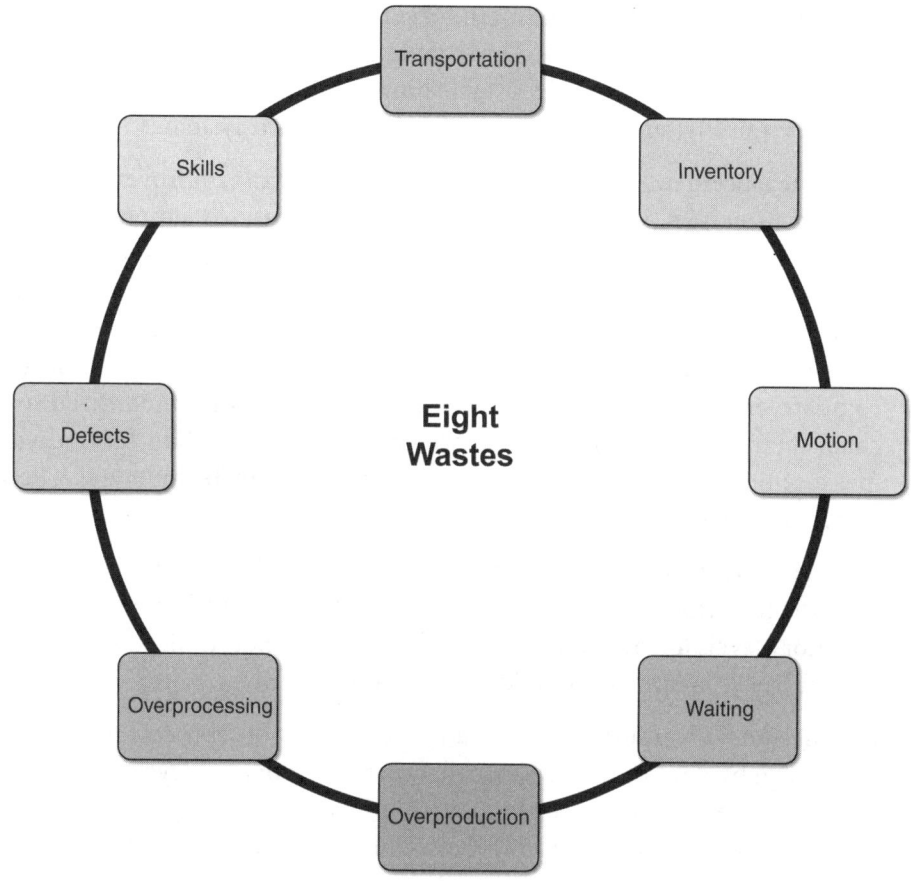

Figure 3.2 The Eight Wastes of Lean

In addition to the eight wastes, there are other wastes to consider in your supply chain and logistics functions, as mentioned by Robert Trent [2008] in his book *End-to-End Lean Management: A Guide to Complete Supply Chain Improvement*:

- **Too many bits and bytes**—The digital age produces less paper—at least in theory—but technology advances guarantee plenty of data. It is important to understand the difference between data and useful information. You can waste a lot of valuable time culling through useless information, such as emails, reports, and analyses.

- **Untapped creativity**—People often take the path of least resistance, but that's not usually the optimal route. Developing creative skills takes time, but organizations can do their part by establishing a culture of Lean thinking, which includes training, support, and a motivational reward system.

- **Poor measurement**—If you can't measure it, you can't improve it. According to Lean principles, you should build your organization's supply chain function to support overall strategies (which can range from responsiveness to low cost), each strategy having its own type of metrics. In today's world, it is easier than ever before to gather measurements, through technology solutions such as bar code scanning and RFID. But it's important to measure, compare, and benchmark the right things. One source of standardized supply chain metrics is the SCOR model, which features more than 150 individual measurements (see Chapter 1, "Lean Supply Chain and Technology: A Perfect Combination").

- **Excessive overhead**—The financial impact of an inefficient supply chain goes right to the bottom line. Such inefficiency can be a result of unnecessarily high inventory levels to cover variability in a process or underutilized assets such as forklift trucks, private fleets, and distribution centers.

- **Overdesign**—Overdesign can result in the waste of overprocessing. Minimize overdesign by developing a Lean, collaborative product lifecycle management process that encompasses a product from inception to disposal of manufactured products. This strategy must also integrate people, data, processes, and business systems.

- **Duplication of effort**—Duplicated effort across sites or geographic locations leads to failure to communicate and understand each other's processes well. Value stream mapping is one way to visualize duplicated effort and come up with solutions that leverage collaboration between organizations.

- **Poor planning**—Many companies do too much reacting and not enough planning. A solid sales and operations planning process at both the grassroots and executive levels can help organizations shift the balance toward planning while monitoring Lean performance metrics.

By focusing on adding value to the customer, you can significantly affect the financial bottom line—and you must not lose sight of that. As a result, identifying and eliminating all kinds of waste is a long-term battle worth fighting in today's ultra-competitive global economy [Myerson, 2015].

In addition to identifying the eight wastes in your supply chain, the following are some other ideas or principles laid out by Martichenko [2013] that are worth considering:

- **Making customer consumption visible to all members of the supply chain**— Flow in the supply chain begins with customer consumption. Visibility to customer consumption for all supply chain partners is critical for acting as the "pacemaker" of the supply chain.

- **Reducing lead time**—Reducing inbound and outbound logistics gets us closer to customer demand, which results in reduced reliance on forecasting, increased flexibility, and reduced waste due to overproduction.

- **Creating level flow**—Leveling the flow of material and information results in a supply chain with significantly less waste at all nodes in the system.

- **Using pull systems**—Pull systems reduce wasteful complexity in planning and overproduction that can occur with computer-based software programs such as material requirements planning (MRP), and they permit visual control of material flow in the supply chain.

- **Increasing velocity and reducing variation**—Fulfilling customer demand through delivery of smaller shipments more frequently increases velocity. This in turn helps reduce inventories and lead times and allows you to more easily adjust delivery to meet actual customer consumption.

- **Collaborating and using process discipline**—When all members of the supply chain can see if they are operating in takt (i.e., demand rate) with customer consumption, they can more easily collaborate to identify problems, determine root causes, and develop appropriate countermeasures.

- **Focusing on total cost of fulfillment**—It's important to make decisions that will meet customer expectations at the lowest possible total cost—no matter where they occur in the supply chain. This means eliminating decisions that benefit only one part of the stream at the expense of others. This challenge can be achieved when all members of the supply chain share in operational and financial benefits when waste is eliminated.

Furthermore, Thompson, Manrodt, and Vitasek [2008] identified six attributes of a Lean supply chain from a survey of Lean practices in the supply chain (see Figure 3.3):

- **Improved demand management**—Lean says to move as much as possible to pull systems, in which products or services are pulled (work initiated, services performed, products delivered) only when requested by the final customer.

To minimize the bullwhip effect described in Chapter 2, "The Growing Importance of the Supply Chain and Technology to Business and Society," a Lean supply chain works to have products pulled through the channel using customer demand from the point of sale in real time.

- **Cost and waste reduction**—In a Lean supply chain, partners have to work together and individually to eliminate wasteful processes and excess inventory across the channel. In general, a reduction in waste will result in a reduction in cost for the supply chain. An organization can identify non-value-added activities from end to end in the supply chain by using value stream mapping.

- **Process (and product) standardization**—Standardization enables continuous, consistent flow of materials, products, and information to occur throughout the supply chain. As businesses become more collaborative, they can see where tasks are duplicated and linked and make improvements. Standardization of the products themselves can also help reduce the number of different components and suppliers, and it can support postponement efforts to reduce inventory levels of finished goods and can also create help suppliers stabilize their own product lines. Today, companies can share intellectual property, metrics, and best practices for many activities, and they can use (standardized) metrics to drive product and process standardization.

- **Industry standardization**—A Lean supply chain needs information to be exchanged and available in a standardized format between all trading partners in order to better communicate and collaborate. Industry product and process standardization can have cost benefits but reduces the proprietary nature of products, making areas like the supply chain even more important from a competitive perspective.

- **Cultural change**—Successfully implementing Lean may require a change in a company's culture. Survey leaders emphasized the importance of Lean and TQM (total quality management) training as part of their new employees' training.

- **Cross-enterprise collaboration**—An enabler of cross-enterprise collaboration is the use of teams. In a Lean supply chain, these teams aren't functionally oriented or internally focused on their organization and tend to have a broader perspective.

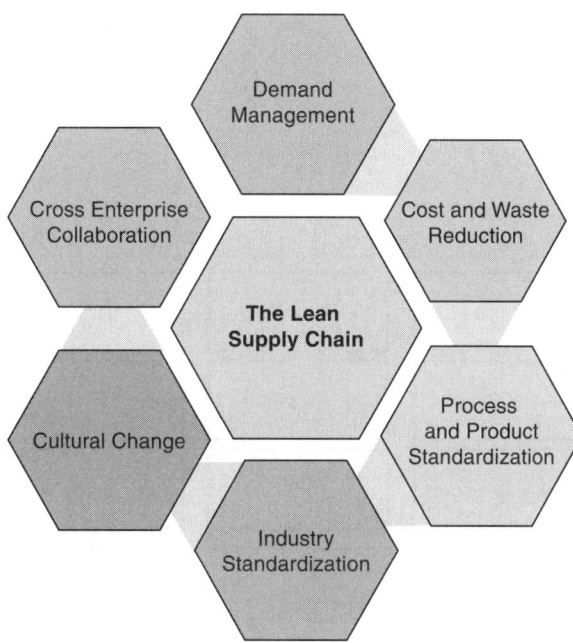

Figure 3.3 Six Attributes of Lean and Effective Supply Chains

In general, Thompson et al. found that Lean supply chain adopters reported improved collaboration, displayed an increased use of standards in processes and materials, reduced SKU counts and inventory levels, and saw a reduction in cost of goods sold compared with non-adopters. A Lean supply chain, therefore, contributes to the bottom line.

Lean Culture and Teamwork

While a variety of tools and methods are used in Lean, the foundation of the technique is a Lean culture that permeates throughout an organization (see Figure 3.4). To be successful, there must be top-down leadership as well as a support (and reward) system to create a bottom-up effect as well. Lean is about teams working on continuously improving processes and activities, so everyone in the organization must be "rowing" in the same direction. This requires executive leadership and also requires making the necessary tools, training, and rewards available to everyone in the organization.

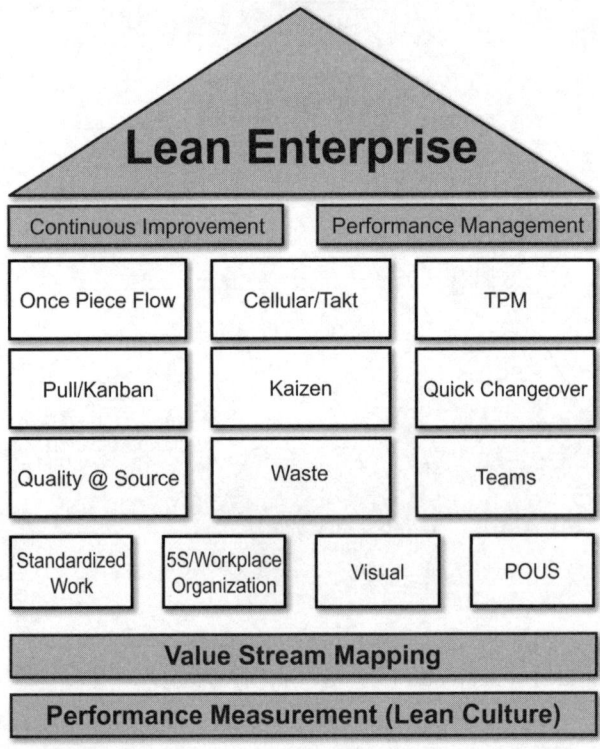

Figure 3.4 House of Lean

Many Lean initiatives are not totally successful, and it is instructive to understand some key success factors (KSFs):

- Train the entire organization and make sure everyone understands the Lean philosophy.
- Ensure that top management actively drives and supports the change with strong leadership.
- Everyone in the organization should commit to make it work.
- Find a good, experienced change agent as the "champion."
- Set a kaizen (improvement events) agenda and communicate it and involve operators through empowered teams.
- To identify value-added and non-value-added activities and an improvement plan as soon as possible, start value stream mapping right away with an important and visible activity.
- Integrate the supporting functions and build internal customer and supplier relationships.

Elements of a Lean Supply Chain Strategy

Before getting into some of the major tools for the Lean supply chain, it's useful to think about, at least on a high level, some of the elements required in establishing a Lean supply chain:

- Identify and eliminate all waste in the supply chain so that only value remains using a collaborative, cross-functional, and interorganizational approach to ensure the smooth flow of materials, products, and information.

- Consider advancements in technology (as described throughout this book) to help enable a Lean supply chain that is flexible and efficient.

- Strive for visibility of customer demand throughout the supply chain to minimize the bullwhip effect, as described earlier.

- Focus on lead time reduction by using various Lean tools to reduce "dock to dock" time, which is a great indicator of how Lean you really are.

- Keep material flowing to ensure speed and flexibility in the supply chain.

- Move the point in your supply chain where you go from "pushing" large quantities to gain economies of scale to demand "pull" of smaller demand-based quantities as far upstream in the supply chain as possible.

As described in the following sections, a variety of tools can help you make sure your supply chain is always striving to attain the critical elements. The Lean toolkit really is an umbrella over a lot of new and old ideas. It can include basic, general concepts such as standardized work, visual workplace, and layout as well as somewhat more complex concepts, such as just-in-time (JIT), kanbans, and work cells. We discuss some of the most important ones and their applications in the supply chain here.

Basic Tools

The following sections describe the basic tools used in Lean.

Standardized Work

Standardized work refers to how work is actually routinely done in the workplace. The idea is to make operations repeatable to ensure consistently high productivity and reduce variability of output, as variability inevitably contributes to waste.

In order to establish standardized work, you must collect and record data on a form. These forms are used to design the process and by operators to make improvements in their own jobs. They can include operations charts that analyze body movement, activity charts that are used to study and improve the utilization of an operator(s), and machines and process charts that use symbols to document movement of people and materials so that non-value-added activities can be identified and eliminated (see Figure 3.5).

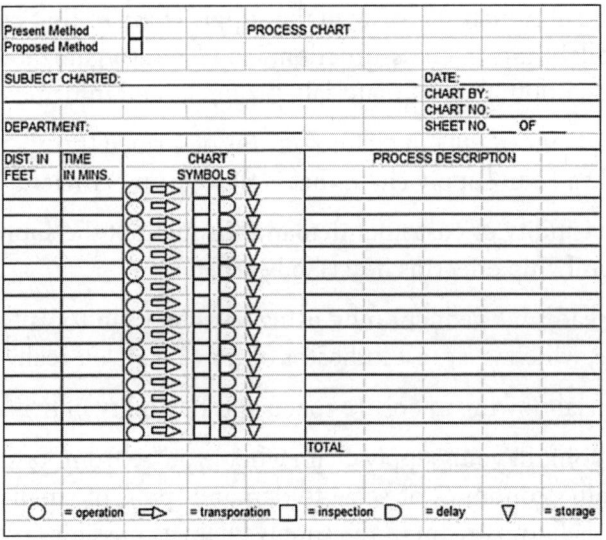

Figure 3.5 Process Chart

Visual Workplace

A *visual workplace*, also known as a *visual factory*, is a Lean concept that emphasizes putting important information at the point of use. Visual systems and devices play an important part in standard work, as do many of the Lean tools we discuss shortly, including 5S, total productive maintenance (TPM), quick changeover, and kanban (pull production).

A visual workplace is also a very important way to sustain these Lean initiatives, because it ensures that Lean improvements remain visible, easily understood, and consistently adhered to well after a kaizen event or process improvement event is finished.

The following are some of the specific areas of a visual workplace:

- **Visual order (or organization)**—Creates a clearly identified place for everything and for everything to be kept in its place, including location identifiers for tools, parts, materials, products, and equipment. Benefits may include reduced inventory, efficient space utilization, enhanced productivity, and reduced operational variability.

- **Visual standards**—Used in standard work, quick changeover, and poka yoke (a mechanism that helps operators avoid mistakes), visual standards are procedures and technical information shown at the point of use to ensure that best practices are followed consistently by all employees and used in setup, operating, inspection, and maintenance instructions (see the example in Figure 3.6). Benefits may include shorter cycle times, improved quality, better safety awareness, and simplified training and scheduling.

- **Visual equipment**—Used in Lean tools such as TPM (which deals with equipment-related waste), standard work, and quick changeovers. Visual equipment helps speed proper setup, ensure correct equipment usage, and easily detect operating abnormalities. This equipment can include operator control labels, gauge indicators, inspection and service labels, and hazard warning labels. Benefits may include faster changeovers, fewer operator errors, simplified autonomous (operator-controlled) maintenance, and fewer defects.

- **Visual production/inventory control (supported by technology)**—Examples are kanbans (demand pull production) and just-in-time (JIT, which involves receiving materials only as they are needed). Visual inventory control involves material flow throughout the supply chain to ensure that the right product is in the right quantity, at the right place, at the right time. It is also used to clearly identify system components and flow of product through a facility. Examples include kanban cards (visual signals indicating when and how much replenishment inventory is required based on downstream pull demand), inventory/bar code labels, RFID tags (electronic tags that automatically identifying and track inventory), and handheld computers. Benefits of visual inventory controls include shorter lead times, reduced inventory levels, improved on-time delivery, and faster troubleshooting.

- **Visual metrics/displays**—These displays are for communicating kaizen schedules, supply chain, and operational metrics, letting employees know of key initiatives, track performance against goals, and recognize efforts and achievements. Examples can include visual dashboards, scoreboards, slogan banners, and kaizen improvement displays. Benefits include better alignment with corporate goals, improved accountability, real-time performance tracking, and increased employee involvement.

■ **Visual safety**—Visual signals ensure that hazards are clearly identified and that employees know how to work safely. Examples can include equipment hazard

			Receiving Form #001
	RECEIVING–LABEL VERIFICATION AND UNJAMMING AT INTAKE		
	Approved by: Draft	**Revision by:** ———	**Version Date: 09/03/10**

PHASE	STEP	COMMENTS	VISUALS
1. 1st Electric Eye-Label Verification	• Inspect LPN and Carlton Label • Press Reset Button	• Numbers must match • If numbers match; if not, take carton off conveyor and see supervisor	  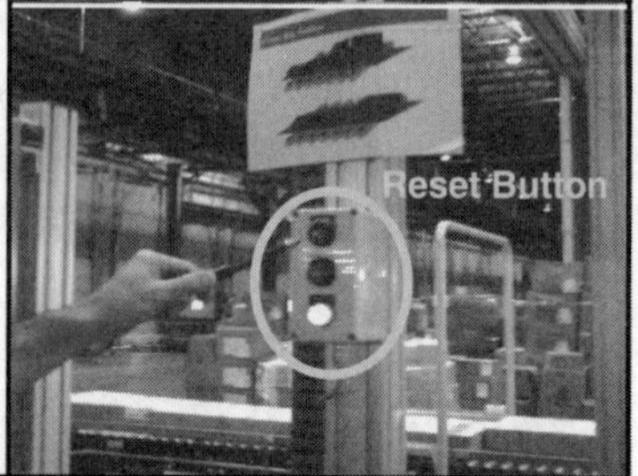
2. 2nd Electric Eye-Un-jam (Clear)	• Un-jam ("Clear") Carton • Push Reset Button		

Figure 3.6 Visual Job Aid

labels, chemical labels, warning signs, and accident prevention tags. Benefits include lower workers' compensation costs, improved employee morale, enhanced regulatory compliance, and reduced downtime.

Layout

Process flow and layout are central to the concept of Lean. Flow patterns arrange process steps in a natural flow order, link process steps to minimize cycle time and travel distance, eliminate areas of congestion, and simulate a continuous flow process by putting internal customers and suppliers next to each other. Figure 3.7 shows an example documenting existing material flow in a distribution center.

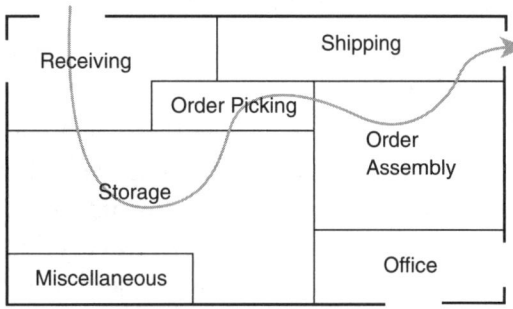

Figure 3.7 Flow Example in a Distribution Center

Facility layout is as important as the technology and materials it contains and has a major impact on business performance. The layout should be reviewed and improved continuously so that waste associated with poor layout can be eliminated or reduced.

There are many reasons the existing production facility layout may not be optimal. For example, an existing facility may limit the choice of a good layout without major reconstruction, and existing layout may not have taken into consideration future expansion or changes to the product mix.

Benefits of improved layout include higher utilization of space, equipment, and people; improved flow of information, materials, or people; improved employee morale; improved customer/client interface; and flexibility.

5S—Workplace Organization

One of the most popular Lean tools is known as *5S*, or *workplace organization*. 5S uses many of the previously mentioned tools, such as standardized work, visual workplace, and layout, to provide a safe, uncluttered, efficient work environment. It is also a great place for a company to start a Lean journey. 5S incorporates the use of standards and discipline and is not just about general housekeeping; it is also about safety, organization, and creating a Lean culture.

5S encompasses the following activities:

- **Sort**—Sort out and separate that which is needed in an area. The motto is "when in doubt, toss it out."
- **Set in order (or straighten)**—Arrange items that are needed so that they are ready and accessible and clearly identify locations for all items so that anyone can find them and return them once a task is completed.
- **Shine**—Clean the workplace and equipment on a regular basis to maintain standards and identify defects.
- **Standardize**—This involves maintaining the first three of the five *S*'s using standard procedures. This might involve going through an area on a regular basis to remove unneeded material and equipment (sort), checking at the end of a shift that materials are in their proper place and checking inventory levels, if applicable, and having a 10-minute clean-up at the end of shift, using a checklist to identify what needs to be done and by whom.
- **Sustain**—Perhaps the hardest *S* of all, sustain requires sticking to the rules to maintain the standards and to continue to improve every day. This involves everything from measurement, audits, rewards, training, and other support, and it really is part of the cultural change that may be required to be successful.

5S provides a solid base for a Lean program. It is generic enough that it is applicable and successful in all industries, in all areas of a company, and with all levels of employees as it allows teams to organize their workplace in a safest and efficient manner.

Advanced Tools

The following sections cover some more advanced tools that are commonly used in Lean programs. While they may require more training and management participation and guidance, they can be very effective when applied correctly and following recommendations.

Value Stream Mapping

Value stream mapping (VSM) is Lean technique that is used to analyze the flow of materials and information currently required to bring a product or service to a consumer. VSM is typically one of the first tools a company should use in creating an overall Lean initiative plan (along with 5S, which enables development of the Lean culture on a broader basis). In some cases, it may be appropriate to precede VSM with a Lean opportunity analysis that identifies the best places to start.

Developing a visual map of the value stream allows everyone to fully understand and agree on how value is produced and where waste occurs. It typically involves the following steps:

1. Identify the target product, product family, or service and determine the takt (i.e., demand time) that is typical at the current time. It is calculated by dividing total work time available by units required (e.g., make one unit and pass one unit every 10 seconds). This determines the pace of the value stream and determines where bottlenecks that limit capacity and create waste may exist.

2. Draw a current state value stream map, showing the current steps, delays, and information flows required to deliver the target product or service (see Figure 3.8). This may be a production flow (raw materials to consumer) or a design flow (concept to launch). There are standard symbols for representing supply chain entities easily found through a search on the Internet.

3. Assess the current state value stream map in terms of creating flow by eliminating waste.

4. Draw a future state value stream map (see Figure 3.9).

5. Implement the future state.

One of the major benefits of VSM is that it helps an organization gains a common understanding of an entire value stream, not just a single area, and learn to identify where the wastes are throughout the organization and extended supply chain. The team can therefore construct a future state of a value stream with less waste and then develop an implementation plan to achieve that future state. The result of a successful value stream mapping process can direct Lean improvements.

Once areas of waste are identified, a variety of advanced Lean tools and techniques (described next) can be used in combination with some of the basic ones already covered.

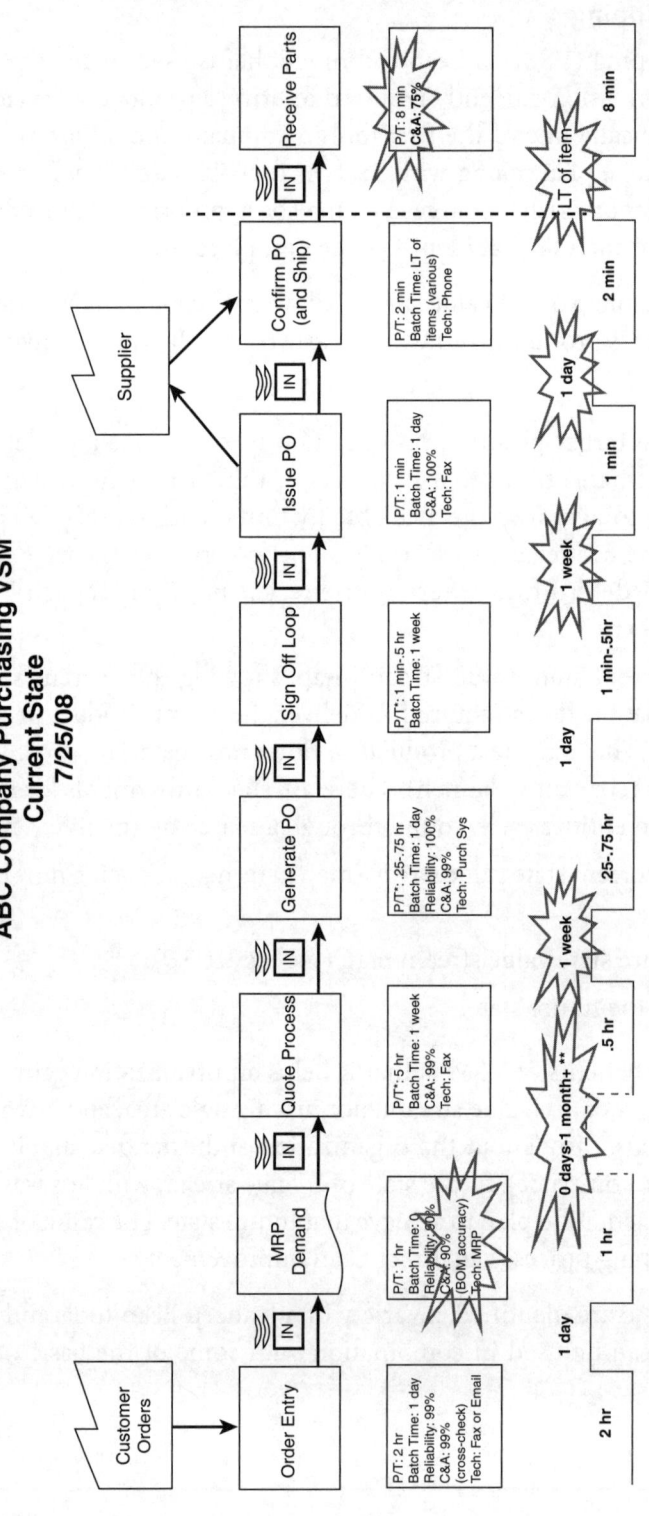

Figure 3.8 Current State Value Stream Map Example

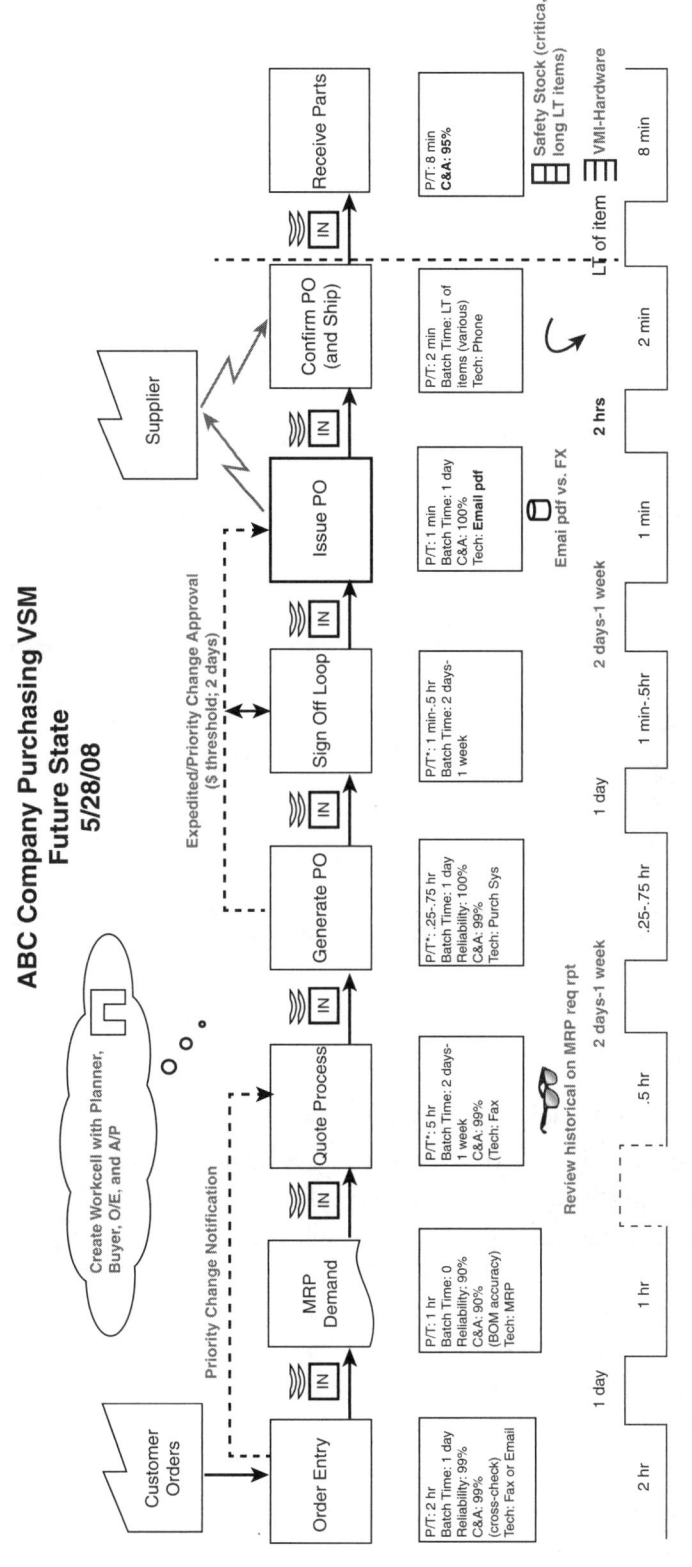

Figure 3.9 Future State Value Stream Map Example

Pull System with Kanban (JIT System)

Just-in-time (*JIT*) is an inventory strategy that companies use to increase efficiency and decrease waste by receiving materials or information only as needed in the process. A *kanban* is a visual signal that is used to trigger an action; it might be a card or even a line on a wall. Typically the action is a request for material or information downstream, in a value stream also referred to as a *pull system*.

In general, a downstream process may only withdraw items in the precise amounts as specified in the kanban, and upstream processes may only send items downstream in the precise amounts and sequences also specified by the kanban. No items are made or moved without a kanban, and typically a kanban card must accompany each item at all times. Of course, defects and incorrect amounts should never be sent to the next downstream process, and the number of kanbans should be monitored carefully to identify changes to downstream demand as well as problems and opportunities for improvement. Figure 3.10 shows an example of a kanban process.

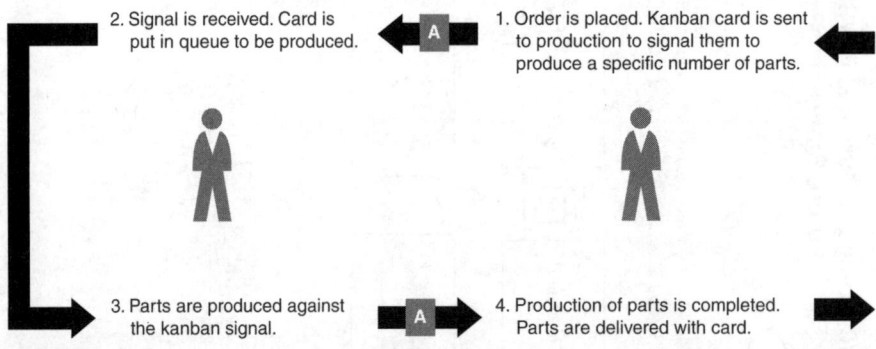

Figure 3.10 Example of Kanban Process with Card

Batch Size Reduction and Quick Changeover

Batch or lot size reduction (ideally one piece, at least in some cases) is an important part of many Lean strategies. Lot size directly affects inventory and scheduling. Other effects are less obvious but equally important, such as throughput time and reduced effects when quality issues arise. Small lots reduce variability in the system and smooth production. They enhance quality, simplify scheduling, reduce inventory, enable kanbans, and encourage continuous improvement. Figure 3.11 illustrates some of the benefits of a JIT process that uses small lot sizes and quick changeovers (for example, lower inventory and shorter lead times).

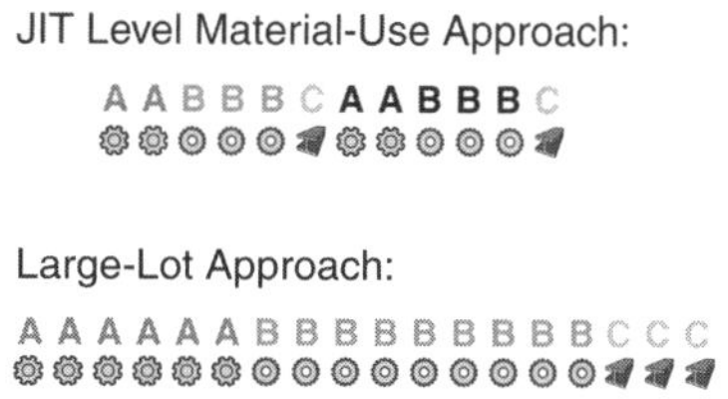

JIT Level Material-Use Approach:

A A B B B C **A A B B B** C
◎◎◎◎◎⬢◎◎◎◎◎⬢

Large-Lot Approach:

A A A A A A B B B B B B B B B B C C C
◎◎◎◎◎◎◎◎◎◎◎◎◎◎◎◎⬢⬢⬢

Figure 3.11 Benefits of Batch Size Reduction

The effects of small lots differ somewhat between make-to-order (MTO) and make-to-stock (MTS) environments but are important in both production strategies.

A changeover is a waste because nothing is produced during that time. In general, there is also a correlation between changeover time (i.e., the full time it takes to go from "last good piece to next good piece," not just equipment setup time) and batch size. That is, the quicker and easier a changeover is, the smaller the batch size can be, helping you to move from a push environment to a pull environment. This is true not only in manufacturing but in the office, in the warehouse, and throughout the supply chain as setups for different activities exist everywhere. For example, batches of orders may sit in an inbox awaiting entry, while the employee does some other activities. Entering orders require a "setup" that includes gathering other needed information and materials, accessing software screens, and so on. Therefore, there is a tendency to batch orders, which slows the throughput time, which leads to orders taking longer to get to the shop floor or warehouse and therefore delays in shipping and billing products.

Many Lean tools, such as layout and standardization, can help shorten changeover time. In many cases it is useful to have a kaizen event focused on a process; the kaizen can be filmed and documented in order to come up with significant improvements.

Work Cells

A *work cell* is an arrangement of resources in a business environment that improves process flow and efficiency and eliminates waste. Work cells are often found in manufacturing and office environments.

A work cell organizes people and machines into groups to focus on single products or product groups. As opposed to a long assembly line requiring a fair amount of labor, a work cell is typically "U" shaped, with equipment that is suited to the family of products or services that flow through it. The labor tends to be cross-trained, empowered, and flexible to ensure the smooth flow of material and information (see Figure 3.12).

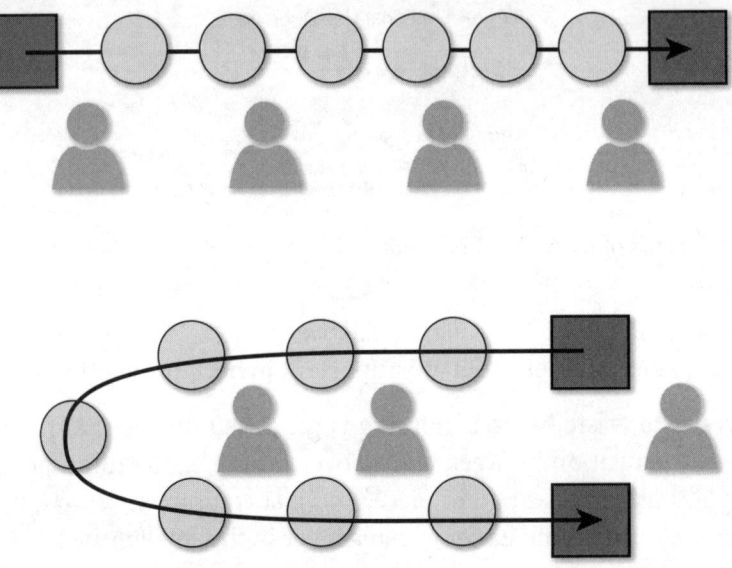

Figure 3.12 Assembly Line Versus Work Cell Layout

Benefits include reduced raw material, work-in-process, and finished goods inventories; better utilization of machinery and equipment; less floor space required; reduced direct labor cost; and increased employee participation.

Total Productive Maintenance (TPM)

Equipment-related waste can have a huge impact on throughput, efficiency, and quality. Yet a surprising number of companies rely on what is known as "breakdown" maintenance rather than doing preventive maintenance.

Much more than just a preventive maintenance program, though, total productive maintenance (TPM) focuses on equipment-related waste and involves a series of methods which ensures that every piece of equipment in a production process is always able to perform its required tasks so that production is never interrupted. It is a comprehensive, team-based, continuous activity that enhances normal equipment maintenance activities and involves every worker, with the goal of keeping equipment producing only good product—as quickly as needed with no unplanned downtime.

TPM involves using machines that are reliable, easy to operate, and easy to maintain, emphasizing total cost of ownership when purchasing machines, so that service and maintenance are included in the cost. It also involves developing preventive maintenance plans that utilize the best practices of operators, maintenance departments, and depot service and training workers to operate and maintain their own machines (also known as "autonomous maintenance").

Using TPM results in many positive outcomes, such as improved equipment performance and availability, increased first-time-through (FTT) quality levels, reduced downtime due to emergencies, increased employee skill levels and knowledge, increased empowerment, improved job satisfaction, and increased safety.

TPM can be applied anywhere equipment is used, not just the shop floor, as distribution centers can't run without forklift trucks and carousels, trucks can't run without proper maintenance, and offices don't perform as well with copiers, printers, and computers that are slow or frequently out of service.

Tools for Identifying and Solving Problems

After identifying waste but before coming up with solutions, it may be critical to find out the source of the waste or variability. This may involve the collection, compilation, and analysis of information to get to the root cause(s) of the problem.

5 Whys

5 Whys is a simple but effective technique for getting to the root cause of a problem. It is attributed to Sakichi Toyoda, one of the fathers of the Japanese industrial revolution, who had a "go and see" philosophy.

To use 5 Whys when a problem occurs, you uncover its source by asking the people who are actually involved in the process "why" at least five times. This usually leads you to the root cause(s) of the problem. For example, say that you find that you shipped the wrong product to a customer. Here's what you might find with 5 Whys:

- *Why?* The wrong item was pulled from inventory.
- *Why?* The item pulled from inventory was mislabeled.
- *Why?* Our supplier mislabeled the item prior to shipping it to our warehouse.
- *Why?* The individual applying labels to our product at the supplier placed the wrong label on the product.
- *Why?* Labels for different orders are preprinted, and it is easy to apply the wrong label.

So in this example, the supplier is the cause of the issue. A variety of things might help eliminate the problem at both the supplier's end (e.g., printing labels to order) and yours (e.g., doing random sampling upon receipt in your warehouse).

Seven Tools of Quality

The seven tools of quality are a set of graphical techniques that help in trouble-shooting issues related to quality. They can also be used in various forms of process improvement to get to the root cause of a problem—quality or otherwise. They are suitable for people with little formal training in statistics and can be used to solve the vast majority of quality-related issues. These are the seven tools:

- **Cause-and-effect diagram (also called a fishbone chart)**—This type of diagram can identify possible causes for an effect or problem and sorts ideas into useful categories.

- **Check sheet**—This form for collecting and analyzing data can be used for a wide variety of purposes.

- **Control chart**—This type of graph can be used to study how a process changes over time within certain, stated parameters.

- **Histogram**—This graph shows frequency distributions, or how often each different value in a set of data occurs.

- **Pareto chart (which expresses the "80/20 rule")**—This type of chart organizes and shows data on a bar graph from greatest type and number of occurrences to least. A Pareto chart illustrates the 80/20 rule, a phenomenon which states that a relatively small number of causes typically generate most of the problems.

- **Scatter diagram**—This type of graph pairs numerical data, one variable on each axis, to look for a relationship (e.g., high absenteeism related to low productivity).

- **Flowchart**—A flowchart shows the steps in a process (i.e., actions that transform an input to an output for the next step). It helps in analyzing a process—but only if it documents the *actual* process used rather than what the process owner *thinks* it is or *wants* it to be. The differences between the actual process and the intended process can provide many ideas for improvement.

Lean, Supply Chain, and Technology

According to the article "How Much Technology Is Needed for Lean Manufacturing?" [*Supply Chain Digest*, 2011], there is much debate over how much technology can help in Lean manufacturing. Most "purists" think that technology plays no role in Lean as it is traditionally known as a very visual and often "pen and pencil" type of tool. However, with the growing complexity of the *global* supply chain, many argue that technology is required to scale Lean in this type of environment.

The reason for the lack of support by traditionalists is that historically, ERP and MRP systems didn't do a great job supporting a Lean philosophy as they were based more on a "push" philosophy—the opposite of Lean.

The following are some of the reasons technology is needed with Lean:

- Over time, many Lean initiatives, which rely on people and their knowledge of Lean practices, tend to deteriorate (or at least go back to their "past" state). Using technology as a foundation to drive Lean processes and information flow helps institutionalize Lean within the factory and supply chain.

- As product mix and production process complexity keep increasing, the low-tech approach to Lean can have a hard time keeping up with things.

- Traditional use of kanban replenishment approaches discussed earlier have to be resized ever more dynamically in complex environments.

- The increased use of outsourcing to make materials and components has made it harder to use kanban cards and visual processes to coordinate component replenishment from external sources. Therefore, more sophisticated supply network communication and collaboration capabilities are needed today.

- There is a greater need to have more real-time data capture systems in order to have success with a Lean philosophy as it is very dependent on timely and accurate information. Technologies such as EDI, bar code scanners, and RFID help ensure greater accuracy and visibility in the internal and extended supply chain.

In fact, an Aberdeen Group report titled "Lean Manufacturing: Five Tips for Reducing Waste in the Supply Chain" [Butcher, 2009] found that organizations that applied proven technology to Lean manufacturing are significantly improving speed, efficiency, and profitability. Aberdeen found that Lean-enabling technology has moved well beyond just electronic kanban and can include Lean-based supplier replenishment extended into the multi-enterprise supply chain, integrating modeling and simulation and value stream mapping to document the production process and value-added activities into more day-to-day applications.

The Aberdeen survey found that best-in-class companies use Lean automation tools such as inventory levels planned based on uncertainty of demand and order management integration with visibility into manufacturing constraints when promising orders. Most are enabling Lean manufacturing practices through demand planning and forecasting systems and segmenting their supply chain using systems for forward-looking inventory targets by customer, location, and so on for production and deployment planning and scheduling [Butcher, 2009].

In order to leverage technology in a Lean environment, it first needs to be a good match for your organization's needs, and it must be reliable and tested. This is the general topic of Chapter 4, "Software and Hardware Sourcing Process and Applications of Supply Chain and Logistics Management Technology."

4

Software and Hardware Sourcing Process and Applications of Supply Chain and Logistics Management Technology

Before delving into the various supply chain processes and discussing where Lean and technology can work together for a business, we need to discuss the identification of requirements and subsequent technology selection and implementation process. Failing to do this in a thorough and thoughtful manner can lead to disaster, resulting in not only less-than-expected benefits but additional costs and delays.

The Procurement Process

In many ways, the software and hardware decision process has much in common with other sourcing decisions for goods and services. Figure 4.1 shows a generic procurement cycle; these cycles may vary depending on the good or service being procured as well as the industry or company involved. The following sections describe the steps in this process.

Identify and Review Requirements

When discussing requirements, procurement activities are often split into two categories, direct and indirect, depending on the consumption purposes of the acquired goods and services (see Figure 4.2).

Direct procurement is production-related procurement, and indirect procurement is non-production-related procurement.

Direct procurement is generally referred to in manufacturing settings only. It encompasses all items that are part of finished products, such as raw materials, components, and parts. Direct procurement, which is a major focus in supply chain management, directly affects the production process of manufacturing firms. It also occurs in retail, and "direct spend" may refer to what is spent on the merchandise being resold.

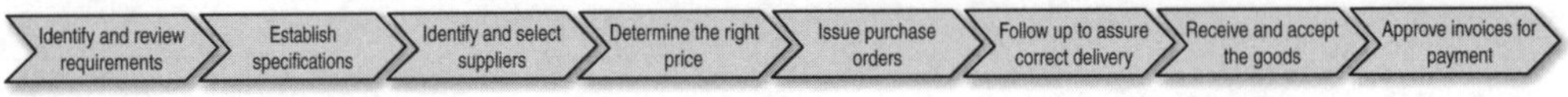

Figure 4.1 The Procurement Process

		TYPES		
		Direct procurement	**Indirect procurement**	
		Raw material and production goods	**Maintenance, repair, and operating supplies**	**Capital goods and services**
FEATURES	**Quantity**	Large	Low	Low
	Frequency	High	Relatively high	Low
	Value	Industry specific	Low	High
	Nature	Operational	Tactical	Strategic
	Examples	Resin in plastics industry	Lubricants, spare parts	Resin and plastic product storage facilities

Figure 4.2 Direct versus indirect procurement

In contrast, indirect procurement activities concern "operating resources" that a company purchases to enable its operations (i.e., maintenance, repair, and operations inventory as well as capital spent on plant, equipment, and technology). It comprises a wide variety of goods and services, from standardized low-value items like office supplies and machine lubricants to complex and costly products and services like heavy equipment, computer software, and hardware and consulting services.

The source for requirements can come from material requirements planning (MRP) systems via planners and purchase requisitions from other users in the organization. (A purchase or material requisition is a document generated by an organization to notify the purchasing department of items it needs to order, the quantity, and the time frame that will be given in the future.) During this step, purchasing reviews paperwork for proper approvals; checks material specifications; verifies quantity, unit of measure, delivery date, and place; and reviews all supplemental information.

Establish Specifications

In order to establish specifications, you must identify quantity, pricing, and functional requirements:

- **Quantity**—In the case of small-volume requirements, you need to find a standard item. In the case of larger-volume requirements, it must be designed for economies of scale to both reduce cost and satisfy functional needs.
- **Price**—This relates to the use of the item and the selling price of the finished product.
- **Functional**—There is a fundamental need to understand what the users expect the item to do. This includes performance and aesthetic expectations (e.g., how smoothly does a hand can opener remove the top of cans, and how ergonomically appealing is its design?).

In general, the description of an item may be by brand or specification. You use brand if the quantity is small or if the item is patented or is requested by a customer. You would use specification if you're looking for very specific physical or chemical makeup, material, or performance specifications.

The source of the specifications themselves can be based on buyer requirements or standards that may be set independently. The buyer setting the specifications can become a long and expensive process, requiring detailed description of parts, finishes, tolerances, and materials used. As a result, the item may be expensive to produce. On the other hand, standards set by government and nongovernmental agencies can be much more straightforward to use as they tend to be widely known and accepted. They tend to lead to production at a lower price and to be more adaptable to customer needs.

Identify and Select Suppliers

The next step in the procurement process is to identify and select suppliers. Typically this involves coming up with a "long list" of suppliers who meet your requirements in general and then whittling the list down to final candidates before selecting the ultimate vendor.

Identification of potential suppliers can come from a variety of sources, including the Internet, catalogs, salespeople, and trade magazine and directories. Once you have identified potential vendors, you issue a request for information (RFI) to them that states a bit about your company and its requirements and requests background on the vendor. It's usually not too difficult to refine the vendors that respond down to a smaller list of candidates (usually 5 to 10), and from there it's best to include a multifunctional team of employees to determine the finalist(s).

Once you have a short list of finalists, you issue a request for quotation (RFQ) or request for proposal (RFP). An RFQ is an invitation to selected suppliers to bid or quote on delivering specific products or services; it includes the specifications of the items/service. The suppliers are requested to return their bids by a set date and time to be considered for selection. Discussions may be held on the bids, in many cases to clarify technical capabilities or to note errors in a proposal. The initial bid does not have to mean the end of the bidding as there may be more than one round.

Vendor Evaluation

The factor rating method, shown in Figure 4.3, is useful in the task of vendor evaluation. You use the factor rating method to identify criteria that need to be considered as part of what you will be buying and assign weights based on the relative importance of each of these factors. You then score how well each supplier compares on each factor and give scores based on score and weight.

While you may have to go beyond using the factor rating method in making a decision, it can get you close enough to help you make a final decision. There are often intangible factors that can come into play, such as personal opinions of executives and prior experience with a vendor.

Criteria	Weights	Scores (1-5)	Weight x Score
Engineering and research capabilities	0.10	4.00	0.40
Production process capability (flexibility/agile)	0.15	5.00	0.75
Delivery capability	0.05	3.50	0.18
Quality and performance	0.20	3.00	0.60
Location	0.05	1.00	0.05
Financial and managerial strength (stability and cost structure)	0.15	5.00	0.75
Information systems capability (e-procurement, ERP)	0.10	2.00	0.20
Reputation (sustainability/ethics)	0.10	5.00	0.50
Total	1.00		3.43

Figure 4.3 Factor Rating Method for Vendor Evaluation

Many factors besides price (and not always the lowest is selected) are important when selecting a supplier, including the following:

- **Technical ability**—As the supplier's product will become part of your product, can the supplier help you develop and make improvements to your product?

- **Manufacturing capability**—Can the vendor consistently meet your stated quality and specifications?

- **Reliability**—Is the vendor reputable and financially stable?
- **After-sales service**—Does the vendor have a solid service organization that offers technical support?
- **Location**—Is the vendor geographically close enough to support fast and consistent delivery and support service when needed?

Determine the Right Price

As mentioned earlier, while price may not be the only determinant in a vendor decision, it certainly contributes greatly to the bottom line as it can be upward of 50% of the cost of goods sold.

Three basic models are used as a basis for pricing:

- **Cost-based**—The supplier makes financials available to the purchaser.
- **Market-based**—The price is based on a published, auction, or indexed price.
- **Competitive bidding**—This is typically used for infrequent purchases but can make establishing a long-term relationship more difficult.

When preparing to negotiate price, preparation is key. On a personal level, if you are buying a house or car, the more research you do, the better idea you have of what is available and what is a "fair" (at least to you) price in the market area. Thanks to the Internet, there are many sources available to get a good idea about what's available and a range of pricing based on recent history. Similarly, in business negotiations, the buyer should to some extent have knowledge of the seller's costs.

Negotiation

For the most part, negotiations are based on the type of product:

- **Commodities**—The price is usually determined by the market.
- **Standard products**—The price is set by catalog listings, and there is usually little room for negotiation (other than volume).
- **Small-value items**—Companies should try to reduce ordering costs or increase volume where possible.
- **Made-to-order items**—Prices are based on quotations from a number of sources, and as a result, prices are negotiated where possible.

Where negotiations are possible, two general types of negotiation can be used: distributive and integrative.

In *distributive bargaining*, the goals of one party are in fundamental, direct conflict with those of another party, resources are fixed and limited, and maximizing the own share of resources is the goal for both parties. So with distributive bargaining, there is usually a "winner" and a "loser."

You need to set a target point and a walk-away point to negotiate a final price that is satisfactory to the buyer. Determining these points may take a good amount of research and judgment. The seller may have a listing or asking price, and you will submit an initial offer or a counteroffer. This type of negotiating usually requires sufficient "clout" to justify lower pricing. Larger companies with multiple locations or business units may have sufficient volume to justify this.

When I was a member of General Electric's corporate sourcing, we were able to leverage over $1 billion per year spent annually on transportation corporation-wide by collecting freight volumes by mode for all of the 100+ GE units to negotiate significant savings. This was accomplished not only by collecting and analyzing the annual spend but also by reducing the number of carriers within each mode to a company-wide group of "core" carriers in order to maximize negotiation power.

Integrative negotiation, on the other hand, is more collaborative, with a goal for a "win–win" conclusion. It involves free flow of information and an attempt to understand the other negotiator's real needs and objectives. This process emphasizes commonalties between the parties and minimizes the differences through a search for solutions that meet the goals and objectives of both sides.

Issue Purchase Orders

At this point, we move from procurement to more of the "day-to-day" supplier scheduling and follow-up, which are really purchasing activities. These activities include execution of the master schedule and MRP to ensure good use of resources, minimize WIP, and provide the desired level of customer service. This usually falls under the auspices of a buyer/planner who works hand in hand with the master scheduler. Buyers/planners are responsible for the control of production activity and the flow of work through the plant and can also be responsible for purchasing, MRP, supplier relationship management, product life cycle and service design, and more. They also coordinate the flow of goods from suppliers.

A purchase order (PO) is used execute the exchange of materials between a buyer and seller. It specifically defines the price, specifications, and terms and conditions of the product or service and any additional obligations for either party. The PO must be delivered by fax, mail, personally, email, or other electronic means.

There are several types of POs, including the following:

- **Discrete order**—This type of PO is used for a single transaction with a supplier, with no assumption that further transactions will occur.

- **Pre-negotiated blanket**—This type of PO made with a supplier contains multiple delivery dates over a period of time, usually with predetermined pricing at oftentimes lower costs as a result of greater volumes (possibly through centralized purchasing and/or the consolidation of suppliers) on a longer-term contract. It is typically used when there is an ongoing need for consumable goods.

- **Pre-negotiated, vendor-managed inventory (VMI)**—With this type of PO, the supplier maintains an inventory of items at the customer's plant, and the customer pays for the inventory when it is replenished (or in some cases when it is consumed). It is usually used for standard, small-value items like maintenance, repair, and operating supplies (MRO) like fasteners and electrical parts.

- **Bid and auction ("e-procurement")**—This type of PO involves the use of online catalogs, exchanges, and auctions to speed up purchasing, reduce costs, and integrate the supply chain. There are many e-commerce sites for industrial equipment and MRO inventory auctions, and they vary in format from catalog (e.g., www.grainger.com, www.chempoint.com) to auction (e.g., www.biditup.com). Websites can be for standard items, or they may be industry specific.

- **Corporate purchase card (pCard)**—A company charge card called a procurement card, or pCard, allows goods and services to be procured without using a traditional purchasing process. There is always some kind of control for each pCard, such as a single-purchase dollar limit or a monthly limit. A pCard holder's activity should be reviewed periodically.

To further enhance the speed and accuracy of transactions, many companies use electronic data interchange (EDI), which is the computer-to-computer exchange of business documents in a standard electronic format between business partners. In the past, EDI transactions either went directly from business to business (in the case of large companies) or through third parties known as value-added networks (VANs). Today, a large portion of EDI transactions flow through the Internet.

Sometimes included in the category of EDI is the use of electronic funds transfer (EFT), which is the electronic exchange or transfer of money from one account to another, within a single financial institution or across multiple institutions, through computer systems. EFT includes e-commerce payment systems, which facilitate the acceptance of electronic payment for online transactions, and it has become increasingly popular as a result of the widespread use of Internet-based shopping and banking.

Follow Up to Assure Correct Delivery

Enterprise resource planning (ERP) software modules such as MRP assume that scheduled dates will be received on time. However, a scheduled delivery date must be monitored and managed in order to identify and avoid possible missed dates in advance where possible. In some cases, delays may be inevitable, and recovery plans must therefore be developed and managed.

It is also critical to have an understanding of the supplier's production process, capacity, and constraints in order to collaboratively resolve problems.

On occasion, expediting is necessary, but it should be done on an exception basis. Supplier performance should be monitored on an ongoing basis. If a particular supplier is consistently being expedited, corrective action should occur.

In many organizations, purchasing may work hand in hand with either their traffic or transportation department or the traffic or transportation departments of vendors (depending on shipping terms as stated in the PO).

Receive and Accept Goods

The key objective at receipt of goods is to ensure that proper physical condition, quantity, documentation, and quality parameters are met. Accomplishing this requires a cross-functional activity among purchasing, receiving, quality control, and finance.

Receiving is technically a non-value-added activity from a customer perspective as it is designed to ensure that everything up to that point has been done properly. The goal is to ensure quality throughout and reduce or eliminate the need for inspection. In many cases, technology such as bar code scanners and handheld computers can be used for automation. Some of the inspection processes can also be reduced or eliminated by having various inspection and certification processes performed by the vendor.

Approve Invoice for Payment

The final step in the procurement process is approving an invoice for payment (see Figure 4.4) according to the terms and conditions of the PO. Typically, the data in the PO is matched with that found on the packing slip that was received and the invoice.

Figure 4.4 Document Flow

Any discrepancies must be reconciled before payment is issued to the vendor. In some cases, small levels of discrepancies can be ignored (e.g., ± 3% or ± $20).

Discounts for early payment should be taken whenever possible, although in a sluggish economy, many customers try to extend payment as long as possible due to cash flow issues.

Software and Hardware Selection

It is important to have a tailored methodology for the successful selection and implementation of technology, which varies somewhat from the previously covered procurement process. While no one method works best, it should include the main phases described in the following sections.

Phase 1—Planning and Budgeting

The initial step in the technology selection process is internal planning and budgeting. You need to set up your project team, get buy-in for the project, and put together a high-level budget.

It is a good idea to have a project advisory or guidance team (or steering committee) to oversee the entire process, beginning with the action plan or a roadmap. This team should include at least one key executive sponsor.

It's not a bad idea to consider using a consultant who has the expertise and knowledge about these steps and different systems and software and hardware products.

Phase 2—Requirements Analysis

Next, you need to put together your requirements document. Make sure it is focused on your requirements (key differentiating criteria) so that you can concentrate on your most important requirements and quickly yet thoroughly evaluate vendors.

During this phase, you should document processes to find potential areas for improvement, using Lean or other techniques. During the subsequent assessment stage, review every step. Examine all steps in each of the business process tasks and make sure each task adds value for the internal or external customer. This may eliminate steps and/or help improve your business's workflow even before a new software implementations or technology upgrade.

Prioritize business needs by determining which features you would like to have and which features you absolutely need to have to determine a complete list of functional requirements.

Make sure to speak with end users to determine more about what could improve their work processes and products and, as a result, identify improvements in cost, time, and user satisfaction.

As there may be a long list of requirements, rank the priorities of these features in terms of what is really required (e.g., required, desirable, and nice to have).

Phase 3—Vendor Research

Now it is time to evaluate your vendor options. This phase focuses on how to start with a long list of vendors and efficiently evaluate them to get to a short list of three or four vendors.

Typically, an RFI is issued first, and it is sent to a long list of qualified vendors. Sometimes an RFP or RFQ is issued if the list isn't too long; if it is, the RFP or RFQ can wait until a little later in the process.

This "winnowing" process can be accomplished by conducting additional research and learning more about the packaged combination of features and benefits available by visiting each service vendor's websites.

Phase 4—Demonstrations

When you have your short list, invite the vendors for demos. Make sure they follow a structured demo "script" so you can see how they will handle your key requirements and so you can compare them in an equal manner. Also look for references that you can contact and possibly visit to "kick the tires."

Phase 5—Final Decision

When you narrow the list to one or two vendors, it's time to do your due diligence and confirm your final decision.

Phase 6—Contract Negotiation

In many cases, software and hardware contracts are written by the vendor. Make sure you negotiate a contract to protect your interests and save money.

Implementation Partner/VAR Selection

A critical part of the selection process that most companies overlook is the selection of a qualified implementation partner or value added reseller (VAR). In certain cases, you need to do this evaluation during the vendor research, and in other cases you select the implementation partner after you select the software product. The selection of the implementation partner can make or break the success of your implementation.

Make or Buy

The first decision in the software/hardware process, at least strategically, is the question of "make or buy," which is the choice between internal production and external sources.

You can use a simple break-even analysis to quickly determine the cost implications of a make or buy decision in the following example.

If a firm can purchase equipment for in-house use for $500,000 and produce requested parts for $20 each (assuming that there is no excess capacity on the current equipment) *or* it can have a supplier produce and ship the part for $30 each, what would be the correct decision: Make (assuming with new equipment) or buy (i.e., outsource production)?

To arrive at the correct decision, a simple break-even point could easily be calculated as follows:

$$\$500,000 + \$20Q = \$30Q$$
$$\$500,000 = \$30Q - \$20Q$$
$$\$500,000 = \$10Q$$
$$50,000 = Q$$

Based on this break-even point of 50,000 units, it would be better for the firm to buy the part from a supplier if demand is less than 50,000 units and purchase the necessary equipment to make the part if demand is greater than 50,000 units.

In terms of the technology decision, in the case of software, much has changed. Just a couple decades ago, there wasn't much packaged software available, so most companies coded and maintained their own applications and even outsourced computer hardware. More recently, there has been a revolution of sorts, starting with the invention of the PC and packaged software. As the year 2000 (i.e., Y2K) approached, there was a fear of old "homegrown" software not accounting for the calendar change (and crashing). In addition, a plethora of ERP and other packaged software systems were then available. These changes greatly reduced the need for in-house programming of entire systems as most organizations migrated toward packaged systems that ran in client/server (and later web-enabled and web-based) environments.

SCM System Costs and Options

The final cost of supply chain management software can be three to five times the cost of the software license as it also includes planning, implementation, training, customization, interfaces, hardware, and configuration of the software. SCM software vendors also typically charge a 15% to 20% annual fee for maintenance and technical support.

A newer alternative to installed software is known as SaaS (Software-as-a-Service), on-demand software, or cloud supply chain software.

Cloud supply chain systems may reduce or eliminate up-front software acquisition costs by offering subscription fees for web-based applications, allowing you to "pay as you go" as fees are based on usage. In this model, there are typically no installation or maintenance costs for the customer.

The major concern of most potential users is security, which may or may not be as big a risk as imagined. Nevertheless, cloud software represents the single highest growth sector in the enterprise software market, and some software vendors are expanding into the cloud by offering some of their SCM modules as SaaS.

According to Gartner, "software as a service (SaaS) SCM offerings showed above-market growth (13 percent in 2012), while perpetual new licenses experienced slower growth of 3.5 percent, as organizations focused on fast implementation at a lower upfront cost" [Gartner, 2013].

"Best-in-Class" Versus Single Integrated Solution

For more specialized types of supply chain applications such as network optimization and forecasting, choosing a "best-in-class" solution may be the way to go as the number of supply chain vendors with a single integrated solution is limited to larger vendors such as SAP and Oracle. In many cases, companies may select one vendor for SCP and another for SCE.

When licensing "best-in-class" software, costs may be greater, as additional interfaces are required with multiple vendor relationships. An application known as an enterprise application interface (EAI) system can reduce some of the integration cost. The benefits of one integrated solution are many, including having a single point of contact, common user interface, and common IT architecture.

Consultants

Three general types of supply chain consultants are involved in the technology selection and implementation process [erpsearch, 2014]:

- **SCM experts or management consultants**—SCM experts help with the planning and modeling.
- **Software vendor consultants**—Consultants employed by the software vendor are application software subject matter experts (SME) who help implement the software.
- **IT consultants**—Information technology (IT) consultants help with infrastructure, interfaces, and custom programming.

The number and mix of consultants in an SCM software implementation project varies depending on the size and scope of the project.

Project Management

Many technology projects are quite complex and involved. It is therefore important to follow good project management principles and techniques.

In general, project management is needed to handle the following characteristics:

- Single unit
- Many related activities
- Difficult production planning and inventory control
- General-purpose equipment
- High labor skills

The use of good project management techniques can ensure the delivery of projects on time and on budget, achievement of project objectives, goal clarity and measurement, coordinated resources, and risk identification and management, among other things. So having a good project management methodology provides a reusable, structured roadmap to managing and executing projects with tools for capturing, analyzing, and improving project work efforts.

Jay Heizer and Barry Render [2013] identify three phases of project management:

1. **Planning**—Goal setting, defining the project, team organization
2. **Scheduling**—Relating people, money, and supplies to specific activities and relating activities to each other
3. **Controlling**—Monitoring resources, costs, quality, and budgets; revising plans and shifting resources to meet time and cost demands

All three phases are critical to the success of projects, especially for technology projects, as many studies have shown a high failure rate. For example, one survey found that 70% were satisfied or extremely satisfied with quality of a project, 48% with the timeliness of the implementation, and 46% with the cost incurred [*InformationWeek*, 2012]. Other surveys and experience from the user side tend to be a fair amount lower in terms of satisfaction with the quality of projects and their results.

Supply Chain Software Market

Today, most companies have implemented at least some components of supply chain systems, such as warehouse management or forecasting. The organizations that have taken an integrated, extended supply chain approach to these systems are the ones that get the greatest benefit.

The general category of supply chain management (SCM) software is also benefiting from the following:

- **Supplier relationship management (SRM) software**—SRM software is a subsystem of SCM software that helps to automate, simplify, and accelerate the procurement-to-pay processes for goods and services.
- **Customer relationship management (CRM) software**—CRM software was originally a standalone system directed at salesforce automation, marketing, and customer service. It is now becoming more integrated with supply chain software such as ERP systems.

- **Product lifecycle management (PLM) software**—PLM software helps companies collaborate and manage the entire lifecycle of a product efficiently and cost-effectively, from ideation, design, and manufacture through service and disposal. It is where applications such as computer-aided design (CAD), computer-aided manufacturing (CAM), computer-aided engineering (CAE), and product data management (PDM) come together.

According to Gartner [2014], the SCM market was $8.3 billion in 2012, which was a 7.1% increase over the prior year. This particular software market is very fragmented: The top 20 vendors account for over half of the market, with the largest 2 vendors, SAP and Oracle, having a 38% combined share, and there are literally hundreds of vendors overall. Driving this large investment has been the need to be more competitive, reduce risk, operate in the global market, and meet various government regulations and industry standards.

SCM systems can be viewed in terms of planning (SCP) and execution (SCE). In general, SCP applications apply algorithms to predict future requirements of various kinds and help to balance supply and demand. SCE software applications usually monitor physical movement and status of goods as well as the management of materials and financial information of all participants in the supply chain.

Supply Chain Planning (SCP)

SCP software vendors address short- to long-term planning and focus on demand, supply, and the balance of demand and supply together, usually in the form of an S&OP process, as described in more detail here:

- **Demand management**—There are three main functions of demand management software: predicting demand, using what-if analysis to create sales plans, and using what-if analysis to shape demand. Forecasts are typically a rolling 24 to 36 months. Because modern supply chain systems have moved toward a demand "pull"-driven model, demand management has moved from purely forecasting to optimizing and shaping demand to some extent.

- **Supply management**—This area helps meet demand with minimal resources at the lowest cost. Software functionality typically found in this area includes supply network planning or optimization (SNP), production scheduling (sometimes referred to as advanced planning systems [APS]), distribution requirements planning (DRP), replenishment, and procurement.

- **Sales and operations planning (S&OP)**—This area facilitates monthly executive planning meetings to tie together sales, operation, and financial plans, along with related tasks to ensure that supply adequately meets demand at the lowest cost. Input is collected from demand, capacity, and financial plans, culminating in a consolidated sales and operational plan.

Supply Chain Execution (SCE)

SCE systems primarily include warehouse management software (WMS) and transportation management software (TMS) and feature planning, scheduling, optimizing, tracking, and performance monitoring:

- **Warehouse management systems (WMS)**—WMS controls the flow of goods through the warehouse and interfaces with the material handling equipment. It also typically includes automated processing of inbound and outbound shipments and the storage of goods. Administrative features can include EDI transaction processing, shipment planning, resource management, and performance tracking.

- **Transportation management systems (TMS)**—A TMS helps manage global transportation needs, including air, sea, ground, and carrier shipments. In terms of transportation acquisition and dispatching, a TMS may also handle the planning, scheduling, and optimizing of shipments. It also provides tracking of vehicles including exception management, constraints, collaborating with partners, and monitoring of freight. Administrative features can include cost allocations, freight auditing, and payment and contract management.

- **Enterprise resource planning (ERP)**—While some may not include ERP systems as SCM tools, a great deal of the functionality is supply chain and logistics related. ERP systems are an extension of an MRP system that tie in all internal processes as well as customers and suppliers. ERP allows for the automation and integration of many business processes, including finance, accounting, human resources, sales and order entry, raw materials, inventory, purchasing, production scheduling, and shipping, resource and production planning, and customer relationship management. An ERP shares common databases and business practices and produces information in real time and coordinates business from supplier evaluation to customer invoicing.

E-businesses must also keep track of and process a tremendous amount of information, and they have therefore realized that much of the information they need to run an e-business (e.g., stock levels at various warehouses, cost of parts, projected shipping dates) can already be found in their ERP system databases. As a result, a significant part of the online efforts of many e-businesses is adding web access to existing ERP systems.

ERP systems have the potential to reduce transaction costs and increase the speed and accuracy of information but can also be expensive and time-consuming to install.

Other Supply Chain Technologies

Other categories of software are often used in the supply chain, including the following:

- **Supply chain event management**—These software applications allow companies to track orders across the supply chain in real time between trading partners, providing managers with a clear picture of how their supply chain is performing. The information provided by these systems allows a company to sense and respond to unanticipated changes to planned supply chain operations. This breed of systems conveys information regarding supply chain processes at a specific event level, such as a hand-off from one supply chain entity to another, the commitment of a product to an order, the movement of a shipment between two logistics network nodes, or the placement of a product into storage.

- **Business intelligence (BI)**—This category is made up of applications, infrastructure, tools, and best practices, providing analysis of information to improve and optimize decisions and performance. BI tools help sort through the vast amount of data that has become available through the continuing adoption of SCM technologies

In addition, there are related tools for supply chain collaboration, data synchronization, and spreadsheets and databases. In fact, many smaller companies today still operate their primary planning functions using spreadsheets and run their day-to-day operations with accounting systems such as QuickBooks and Peachtree rather than spend the resources on a full-blown ERP system.

Current and Future Trends in Supply Chain Software

Short-Term Supply Chain Technology Trends

The 2010 annual Gartner supply chain study [Gilmore, 2010] looked at supply chain application areas and where companies say they stand in terms of adoption:

- The top application area fully implemented (not including ERP systems) was warehouse management systems, which were fully deployed by only 39% of respondents. That was followed by supply chain planning (32%), sales and operations planning (29%), and transportation management systems (28%).

- The top three obstacles to achieving their company's supply chain goals were forecast accuracy/demand variability (59%), supply chain network complexity (42%), and lack of internal cross-functional collaboration and visibility (39%).

- According to the study, the investment priorities when it came to supply chain technology were "improving planning processes" (20%), followed by "aligning corporate and supply chain strategies" and "improving supply chain visibility" (both at 11%).

Interestingly, in the 2013 study [Gilmore, 2013], partially due to continued sluggish economic growth, the focus on using the supply chain to drive business growth was the top priority of companies surveyed, with customer service in the second spot.

The obstacles to reaching supply chain goals, however, were similar to the ones mentioned previously from the 2010 survey.

Emerging Supply Chain Technology Trends

While the types of software applications in supply chain probably won't drastically change, the methods for gathering data and for using and sharing applications will. The following have been some of the major areas of innovation most recently:

- **Cloud computing**—As mentioned previously, this method, also referred to as SaaS (Software-as–a-Service), delivers a single application through a browser to thousands of customers "on demand," avoiding costly licensing, implementation, and maintenance costs. It allows companies to focus on their core competencies while allowing a third party to manage technical elements. Salesforce.com is probably the best-known example among enterprise applications, but cloud computing is also common for HR and ERP applications, as well as some "desktop" applications such as Google Apps.

- **Mobile computing**—Supply chain execution and event management is becoming more mobile, with basic visibility and traceability available on smart phones and other mobile devices.

- **3PLs providing technology**—Cost is a big driver, and 3PLs can offer economies of scale, especially for small and mid-size companies.

- **Radio frequency identification (RFID)**—RFID is an automatic identification method using electronic tags that have a microchip and printed antenna. They can be put together in a variety of forms (e.g., label or imbedded in between the cardboard layers in a carton or product packaging).

The 2010 Gartner survey [Gilmore, 2010] showed 51% of the companies surveyed not doing anything with RFID. To become more widespread, RFID costs will need to continue to decline until it is more economically feasible. In addition, equipment issues such as reader range, sensitivity, and durability need to improve.

An example of the use of RFID is with Intel, the global semiconductor manufacturer that needs to know where its product is at all times. Intel is embarking on a joint effort with DHL, utilizing sensors to monitor the condition of containers as they move around the world. Intel has another project, in Costa Rica, using RFID technology to minimize handheld scanning of inbound and outbound shipments. It has achieved labor savings of greater than 18% as a result of faster processing and has also eliminated steps in these processes [Harrington, 2007].

In addition, emerging supply chain technology being developed now will have a major impact in the near future, including the following:

- **Multi-enterprise visibility systems**—These systems provide a comprehensive and timely view of processes, solutions, and metrics across the entire value chain. When implementing collaborative programs such as VMI, outsourcing, or just-in-time, it is important to also implement the infrastructure or processes necessary to manage inventory in this extended supply chain. This type of emerging solution offers a 360-degree view of supply chain events.

- **People-enabling software**—This technology empowers people to analyze, find, use, collaborate, and share data to maximize efficiency and workflow. ERP and other enterprise software solutions help enable and automate business processes, but they only alert users when problems occur and don't help solve the problems themselves. Technology companies are coming out with productivity tools that enable people to combine unstructured information and business processes with the structured business processes that ERP applications provide. This type of technology platform would enable someone to handle multiple alerts to a smart phone, for example, empowering them to put fires out on the spot by connecting customer, manufacturer/distributor, and supplier systems on a mobile device.

- **Execution-driven planning solutions**—These tools utilize data from current executed processes to drive future planning and forecasting. Over the years, many companies have had a disconnect between planning and execution. These systems will use information about the current state of a business to help drive planning decisions for the next planning period in real time. This will give users the ability to consolidate and aggregate massive amounts of data in meaningful ways by applying machine learning techniques to data-mining algorithms to detect data trends. Businesses will therefore have a chance to respond to problems and take advantage of opportunities much faster than before.

- **Human supply chain technology**—These solutions apply supply chain technology to the management of human resources (i.e., the labor supply chain) and allow companies to standardize job descriptions, capture spend and labor rates, and improve and manage their labor hiring practices.

At this point, you should have a good fundamental understand of the supply chain and logistics function, Lean concepts, and the general types of technology involved. We will now look in more detail at specific software and technology and how they can enable a Lean supply chain using the SCOR model in Part II, "Plan."

PART II

Plan

Supply Chain Network Optimization

Before delving into types of technology used for planning in the supply chain, it is important to note that some planning tools are more strategic in nature, while others are more tactical and operational. We will briefly discuss planning levels before moving on to supply chain network optimization tools, which are strategic in nature.

Strategic Planning Level

In general, a strategy is a road map for the entire supply chain process. Strategic supply chain decisions are usually the first step in developing a good process. The supply chain strategy should align with the organization's overall mission and strategy, especially in terms of cost, responsiveness, differentiation, and quality.

Issues addressed at the strategic level include the following:

- Choosing the site and purpose of business facilities
- Creating a network of reliable suppliers, transporters, and logistics handlers
- Planning long-term improvements and innovations to meet client demands
- Planning for inventory and product management throughout its lifecycle
- Implementing information technology programs and systems to make the process more effective

Tactical Planning Level

Shorter-term decisions involving the supply chain occur at the tactical level, where processes are defined. Tactical decisions play a big role in controlling costs and minimizing risks where the focus is on customer demands and achieving maximum customer value.

Contingency planning, which helps guard against issues related to unpredictable changes in distribution operations, can be used both strategically and tactically in an organization as the supply chain is greatly impacted by globalization and its inherent logistical complexity. There is risk beyond just the demand and supply variability, limited capacity, and quality issues that domestic companies have traditionally faced, including other trends such as greater customer expectations, global competition, longer and more complex supply chains, increased product variety with shorter life-cycles, security, and political and currency risks. As a result, it is becoming increasingly important for global supply chain managers to be aware of the relevant risk factors and build in suitable mitigation strategies.

Tactical concerns may include the following:

- Procurement contracts for required materials and services
- Production schedules and guidelines to achieve quality, safety, and quantity standards
- Transportation and warehousing solutions, including outsourcing and third-party logistics options
- Inventory logistics, including storage and finished goods distribution
- Achievement of best practices in comparison to competitors

Operational Planning Level

The operational level of supply chain management is where "the rubber meets the road" so to speak. This is where day-to-day processes, decision making, and short-term scheduling take place. It is critical not to jump straight into operational management without focusing on the strategy and tactical levels. Effective operational level processes result from strong strategical and tactical planning.

Operational-level management issues may include the following:

- Daily and weekly forecasting to satisfy demand
- Production operations, including scheduling and detailed management of work in progress
- Monitoring of logistics activity for contract and order fulfillment
- Settling of damages or losses with suppliers, vendors, and clients
- Management of incoming and outgoing materials and products and on-hand inventories [Po, 2012]

This chapter covers the technology used in more strategic decisions and then the following chapters move to some applications used in more tactical decisions. In some cases, such as forecasting, technology can be used to support all three levels of supply chain planning.

Importance of the Supply Chain Network

Location is a strategic, long-term decision that is not easily changed in the short term and applies to raw material sourcing, manufacturing, distribution, and retail. In order to remain competitive in today's global economy, the efficient movement of goods from raw material sites to processing facilities, manufacturers, distributors, retailers, and customers is critical. Unlike transportation and inventory decisions, location decisions tend to be less flexible as many of the costs are fixed in the short term.

Strategically, for a manufacturer the major goal or priority of the location decision is to minimize cost, while retailers look to maximize revenue where possible. Picking the wrong supplier, manufacturing, or distribution location can have a long-term impact on the total cost of a product. This decision can be heavily influenced by transportation costs as they can average 3% to 5% of sales, with warehousing costs being 1% to 2% on average, historically.

With that in mind, then, the goal of supply chain network design is to determine the best location of facilities within the supply chain, while determining the capacity of these facilities, determining how to source demand through the network, and selecting modes of transportation in a manner that provides the required level of customer service at the lowest cost.

The Location Decision and Its Impact on Value

As with many other aspects of the supply chain, there are many trade-offs to consider in the location decision. In terms of supply chain network design, there is the major trade-off of cost versus service (i.e., the level of investment spent on inventory, transportation, and distribution in relation to the service level offered to customers).

From the customer's perspective, service may be viewed in a variety of ways, including the following:

- **Lead time**—The amount of time for a customer to receive an order
- **Product variety**—The number of different products offered by a distribution network

- **Product availability**—The likely hood of a product being in stock when the customer places an order

- **Customer experience**—Has many dimensions, including how easy it is for customers to place and receive orders as well as how much the experience is customized

- **Time to market**—The time it takes to develop new products and bring them to market

- **Order visibility**—The ability of customers to track orders from time of placement to delivery

- **Return-ability**—The ease with which a customer can return merchandise and the efficiency of the network to handle these returns

In general, companies selling to customers who can handle a relatively long response time may require only a few locations that are far from the customer. In these cases, companies may concentrate on increasing the capacity of each location. On the other hand, companies that sell to customers who are looking for short response times and maybe even picking up product with their own vehicles, need to locate facilities close to them. These companies typically have many facilities, each with relatively low capacity. So the trade-off here is that faster response times required to meet customer demand increase the number of facilities required in the network and, conversely, a decrease in the response time customer's desire increases the number of facilities required in the network (see Figure 5.1).

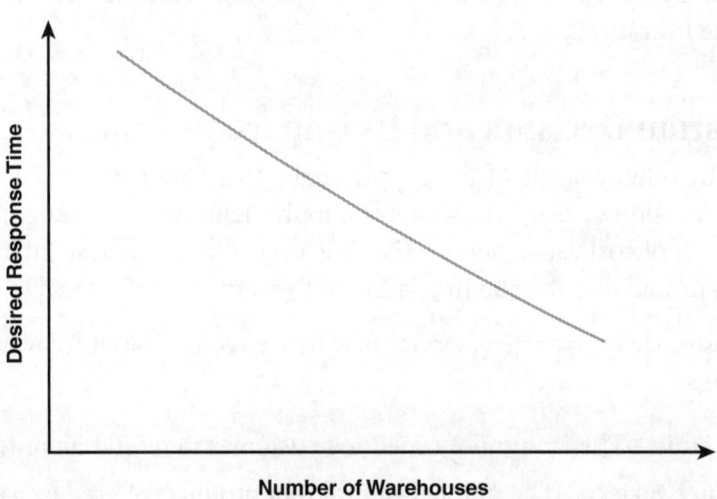

Figure 5.1 Relationship Between Number of Warehouses and Response Time

Changing the distribution network design affects other supply chain costs (see Figure 5.2), including the following:

- **Inventories**—The more locations there are, the harder it is to accurately forecast demand as there are smaller and smaller demand groupings, making the target smaller and harder to hit. As a result, safety stock requirements go up almost exponentially (i.e., the "square root" rule, which states that average inventory increases proportionally to the square root of the number of locations in which inventory is held).

- **Transportation**—The ideal is a "long in and short out" to gain economies in transportation on the inbound end (i.e., full truckloads versus less-than-truckload), but this of course can reach a point of diminishing returns as it increases inventory and warehouse operating costs.

- **Facilities and handling (i.e., warehouse operations)**—Certain economies of scale are gained by operating fewer warehouses, whether company owned or outsourced by consolidating volume. This can result in lower unit handling and storage costs with fewer facilities and must be analyzed thoroughly. On the other hand, the use of local public distribution centers may enable a company to have its products combined with other companies' products to gain some local transportation savings.

In addition, as the number of distribution facilities increases, the amount of information to manage increases. This can be somewhat mitigated by having efficient and integrated information technology systems.

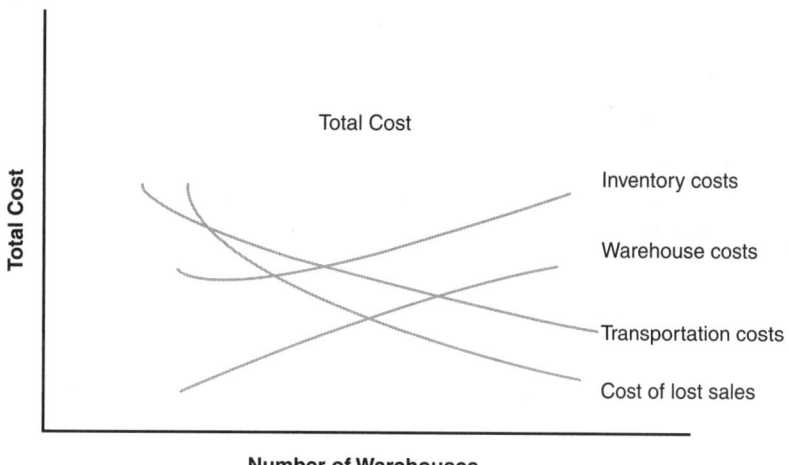

Figure 5.2 The Number of Warehouses and the Impact on Cost

How and Why Network Optimization Technology Can Help Enable a Lean Supply Chain

As customer requirements become more complex in today's demand-driven, omni-channel environment, supply chain optimization studies are the foundation for some of the most successful companies' logistics and fulfillment operations. As a result, supply chain networks need to be frequently reevaluated, as happens with the Lean concept of continuous improvement, as organizations can strategically create value and enable profitable growth in new and existing markets by optimizing supply chain performance.

From a Lean and agile supply chain perspective, an optimally designed supply chain can significantly improve margins, support expansion into new markets, enhance the customer experience, and reduce operating costs. In many ways, the process of supply chain network optimization can help you achieve more value and less waste through keeping lower inventories, maintaining the right stock levels, and choosing the right transportation modes and warehousing strategies throughout your network.

Without an optimized supply chain network, it becomes very difficult to execute on a Lean strategy. For example, if your business ships small orders and has grown in the Southwest United States, but your finished goods distribution centers are located in the Midwest and Northeast, you know in general that it might make sense to consider opening a new distribution center in the Southwest. This type of analysis can not only help you determine the precise size, location, and market to serve but also the savings and improved service levels of various what-if scenarios.

As supply chain network optimization studies can be fairly technical—requiring specialized software and often consulting advice with a price tag as high as $100,000— many small to medium-sized enterprises (SMEs) tend to either not do this type of study at all, put it off, or just use a rough guess-timate. As a result, they may leave significant money on the table, as well as a variety of wastes in their internal and extended supply chains.

Network Optimization Technology

Network optimization software evaluates the total supply chain from manufacturing and suppliers through warehouses and distribution centers all the way to distributors and end customers. The typical tools used to perform supply chain network design are based on quantitative modeling and optimization, which refers to the selection of a best element (with regard to some criteria) from some set of available alternatives.

The use of models enables you to evaluate complex relationships and trade-offs of the overall system by connecting large numbers of variables in a framework that makes it easier to define relationships. Quantitative tools to perform this type of analysis also make the translation of an operational strategy into a financial business case much more straightforward.

Types of models used in network design include spreadsheets, regression and statistical analysis, simulation, linear programming, mixed integer linear programming and expert programs/heuristics. In general, the trade-offs between these different approaches involve speed versus complexity and the need to achieve "good enough" versus optimal solutions. Typical commercial network optimization solutions look at long-term demand forecasts, supply chain facilities and capacities, lead times, and fixed and variable costs to identify the most cost-effective supply chain network. The deliverables usually include where to locate facilities, the size of these facilities, what modes of transportation to use, and long-term sourcing decisions. As the name implies, network optimization tools seek to optimize performance across an entire supply chain network.

This breed of tool can be used either as an integrated part of a suite of advanced planning and scheduling (APS) tools or as a standalone application to analyze only supply chain design decisions. In many cases, as this type of analysis is performed relatively infrequently (maybe every one to three years) and requires some specific expertise, companies may choose to bring in consultants who may have their own software.

Integrated approaches can offer a distinct advantage in that the network design system may use modeling criteria similar to the modeling used for day-to-day supply chain planning activities. As a result, integrated models are often better maintained. However, this approach, because it involves the integration between the various systems, may take longer to obtain results than would be achieved using dedicated models based on more manually massaged data.

As network optimization tools model the entire supply chain, they require a lot of information, including forecasts, product and facility information, manufacturing, storage and distribution rates, capacities, a variety of costs, and objectives such as inventory turns, services levels, and so on.

Other planning tools, which we will discuss later in this book, typically model a supply chain at the SKU/location level, but this is not always the case for network optimization models. Many models such as this that are used to evaluate strategic decisions utilize somewhat aggregated data. Aggregated data reduce the challenges of adding new plants, distribution centers, or demand for what-if scenarios; aggregated data also reduce model size and runtimes, allowing what-if scenarios to be more easily solved and evaluated [Spinnaker, 2015].

Technology Options

A variety of network optimization solutions are available today. They range from standalone systems to modules of larger supply chain systems and can be installed or can be on-demand cloud software systems.

JDA, JD Edwards (Oracle), SAP, and Logility all have network optimization modules that are integrated with their other supply chain planning and execution modules. Other systems, like IBM ILOG LogicNet Plus XE and Logistix Solutions, for example (see Figure 5.3), offer standalone systems (on demand, in the case of Logistics Solutions). They come with lower upfront costs, but because they aren't integrated with other supply chain planning and execution modules, they may be more data intensive.

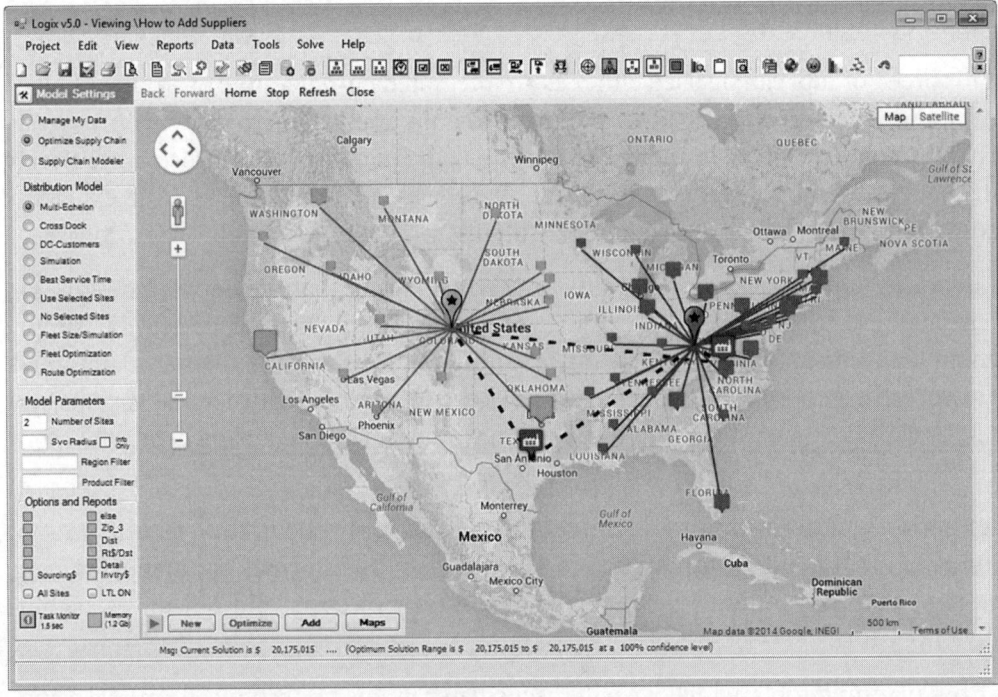

Figure 5.3 Logix Example (Copyright © Logistix Solutions, LLC, Map Data © 2014 Google INEGI)

Supply Chain Network Optimization Technology Case Studies

The following sections provide some actual examples of companies that have used supply chain network optimization technology to reduce waste and improve the efficiency in their extended supply chains.

Case 1: North American Industrial Manufacturer Optimizes Its Distribution Network

Challenge

In a not uncommon scenario, a leading industrial manufacturer had a complex manufacturing network and a large number of distribution centers across North America. However, it didn't have an effective method to evaluate the options available and make good, strategic business decisions with regard to the correct number and location of facilities. The company competed with a small number of known competitors and wanted to model its supply chain and perform scenario analyses to evaluate how customers' sourcing and buying behavior might change in different industry-wide supply and demand situations.

Approach

This client hired Spinnaker (www.spinnakermgmt.com), a supply chain consulting organization, to help design and implement a network design analysis tool that could be used to model and analyze different what-if scenarios, using Oracle's Strategic Network Optimization (SNO) supply chain modeling software. This solution utilized data from the client's production systems as well as information for what-if scenarios that could be created and maintained outside the production systems.

The network design model provided a tool to be used to evaluate changes in the client's supply chain and market environment, including addition or removal of warehouses/production facilities and changes in manufacturing capacities at existing facilities, demand by product line and/or geography, cost, and current and potential competition sites and capacity.

Results

Supply chain network analysis used to take this company either weeks to complete manually or didn't occur at all because of the effort required. After the project, the client was able to evaluate more than 50 what-if scenarios, resulting in significant supply chain design changes, including the opening and closing of facilities. The modeling effort has led to supply chain cost reductions and generated more than 20 times return on investment while providing an increase in customer service [Spinnaker, 2015].

Case 2: Semiconductor Manufacturer Re-Designs Its Global Supply Chain Network

Challenge

A client of Establish Inc. (www.establishinc.com), a supply chain management consulting firm that was a global manufacturer of semiconductor products, was utilizing two main global distribution warehouses in the Asia-Pacific region. It had some space constraints at its facilities, as well as some business model and operational changes that were affecting the logistics flow.

Approach

The first step in evaluating the network was to determine whether the current network was optimal with the changes that had taken place and, if not, recommend what the optimal network should look like.

Establish utilized its significant global modeling expertise as well as CAST software (a global supply chain modeling software from Barloworld Optimus) to design the optimal network.

The client had more than 10,000 different products, delivery requirements of 48 to 72 hours, and diverse customers throughout North America, Europe, Asia, and the Middle East. So the model needed to be both flexible and precise to identify the optimal network configuration.

The Establish team gathered data from multiple sources on different continents to document current supply chain flows and costs. A baseline model was developed for an accurate representation of the current supply chain network. The baseline was used as the basis for comparison with all the various model runs.

Results

The model was run many times, with changes in physical locations and number of locations, transportation modes, delivery requirements, potential customer changes, and product characteristic variations.

In the end, an alternative configuration was found that was more cost-effective and offered similar delivery service levels; it was a combination of existing locations with the movement and consolidation of other locations to produce an optimal configuration. The client was expected to have a 20% savings in total supply chain costs with the new configuration [Establish Inc., 2015].

Even with an optimized supply chain network, the absence of a good demand forecasting process enabled by technology can result in waste throughout the supply chain, which is the topic of Chapter 6, "Demand Forecasting Systems."

6

Demand Forecasting Systems

U p until 25 years ago or so, forecasting was kind of the "red-headed stepchild" that no one wanted to take ownership of. I believe that had a lot to do with the fact that as forecasting basically tries to predict the future, there was (and still is, to some extent) an air of mystery surrounding it. In many cases this led to separate and somewhat disconnected processes and systems for forecasting.

Marketing and sales would forecast dollars by product (sometimes by customer) and brand at least at an aggregate for budgeting and planning purposes, but these numbers tended to be updated only quarterly at most, thus becoming "stale," and were not truly statistically based (and in the case of sales, they were highly opinion based and potentially skewed toward bonus goals).

Manufacturing and supply chain needed to run the day-to-day business and therefore needed shorter-term, accurate (i.e., at least partially statistically based), current, SKU-based (i.e., an individual item at a stock-keeping location) forecasts to drive inventory deployment, production planning and scheduling, procurement of materials, staffing, and equipment requirements. So in many cases, they developed their own sets of forecasts. In many cases, this led to a two-number system that caused financial, inventory, and customer service issues.

In the late 1980s, with the help of forecasting software systems such as American Software (which later became Logility), companies were able to tie it all together with one-number systems that allowed for a kind of "pyramid" approach to forecasting. Statistical SKU forecasts (with qualitative adjustments) could be aggregated to higher levels, such as brand or family, discussed among departments, and possibly adjusted at that level; the adjustments could then be prorated back to the SKU level (and in a variety of units of measure), thus resulting in a consensus, one-number system.

Starting in the late 1980s, organizations began to create entire forecasting functions and departments with trained, experienced forecasters (aka demand/forecast analysts, demand planners) who were able to blend art and science (i.e., qualitative and quantitative forecasting methods) to develop more accurate, consensus forecasts. More often than not, this function can be found under the supply chain and/or operations organization, although in some cases, it can still be found under the sales organization.

In the past several decades, companies have realized the importance of forecasting to all levels of business planning. If you think about it, most aspects of business strategy, planning, and operation are based on some kind of forecast.

A Lean Approach to Forecasting

Forecasting, perhaps more than other process, needs to be Lean and efficient, or it can drive waste throughout an organization in the short, medium, and long terms. Consider these examples:

- **Strategically**—The accuracy (or inaccuracy) of a forecast can wreak havoc on a supply chain network in terms of locations, functions, layouts, inventory, and operating costs and ultimately customer service levels.

- **Tactically**—Forecasting can affect budgeting and production and deployment planning.

- **Operationally**—Forecasting can affect purchasing, job scheduling, workforce levels, job assignments, and production levels.

Ultimately, forecast inaccuracy (at any level) can either lead to waste to the customer in terms of short, late, or no delivery (and hence lost sales); larger amounts of safety stock inventory (and costs) to compensate for forecast inaccuracy; and potentially any of the other eight wastes, in addition to excess inventory, as described in Chapter 3, "Lean Concepts and Their Applications in the Supply Chain."

The first step in Lean forecasting is to ensure that there is a robust forecasting process in place. Next, you need to look for opportunities to improve the process. It should focus on minimizing the use of company resources to maximize customer value by creating meaningful, accurate forecasts as efficiently as possible.

Typical Forecasting Process

The following are the typical steps in a best-practice forecasting process:

1. **Determine the use of the forecast**—Forecasts can be used as a "driver" for many business decisions, including budgeting, capital investments and improvements, production, and inventory deployment. Knowing this up front helps determine not only data and software requirements but also best-practice processes.

2. **Select the items to be forecasted**—Many items, especially those most important to the success of an organization, rely on accurate forecasts. Those items should get the bulk of the resources to have forecasts as accurate as possible. Other, slower-moving items with lower profit margins may be best managed with simpler methods of forecasting or even rely on min/max inventory policies alone.

3. **Determine the time horizon of the forecast**—The time horizon depends on the intended purpose of the forecast. If a forecast is to be used more for long-term capacity and new product planning decisions, then it may need to go out a number of years and be in annual planning periods, or "time buckets." If a forecast is used to drive short-to medium-term processes such as master production scheduling and short-term planning, an emphasis is placed on

the short term, and the data are typically in smaller planning periods, such as months, weeks, or even days.

4. **Select the forecasting model(s)**—Some items, such as slow-moving, low-margin items, can utilize simple forecasting methods such as a moving average. On the other hand, a faster-moving item with seasonality might require a range of more sophisticated statistical methods such as regression. Whatever combination of statistical methods are used might be adjusted based on qualitative methods, determined by the forecaster's expertise or intuition, as well as information from elsewhere, such as sales or the customer.

5. **Gather the data needed to make the forecast**—The system must be capable of gathering recent demand history in an efficient manner as many companies have many thousands of SKUs to forecast. This is usually done via integration with other modules or systems that are the source of this transactional data.

6. **Make or generate the forecast**—Forecasts need to be re-generated with the more current information, using the various models and methods discussed in step 4.

7. **Validate and implement results**—Forecast accuracy targets should be set based on some realistic and meaningful basis. This can include having tighter targets for "A" items (i.e., the "critical few" items that generate most of a typical business's sales and profits) and lower-accuracy targets for the many "C" items (i.e., the majority of items, which individually don't generate much in the way of revenue or profits) or other criteria, such as high shortage or holding costs, anticipated engineering changes, or delivery or quality problems. Once targets are set, it is important to not only measure forecast accuracy but also determine the cause(s) of higher-than-anticipated errors.

Lean Forecasting Process

Kahn and Mello [2004–2005] identified five steps in Lean forecasting:

1. Specify the value that channel partners get from forecasting. Primarily this involves getting the product delivered to them at the right time, quantity, place, and price as well as helping to reduce their supply chain costs through reduced inventory.

2. Identify the value stream and focus on eliminating waste such as excess data collection and reporting, long queue times for information, over-analysis of data, too many or incorrect people involved in the forecasting process, and high system costs.

3. Create flow especially by reducing the time between receiving information and making decisions by focusing on reducing the time between creating a baseline forecast, making adjustments, and final approval.

4. Facilitate pull by creating a procedure to initiate when a forecast should be made. Pull can also refer to which items will be forecasted and which will be managed more through inventory control policies, such as relying more on reorder points with safety stock or kanbans.

5. Strive for perfection to create the most meaningful forecasts from the customer perspective.

They also pointed out a framework of focal elements to assist in Lean forecasting initiatives:

- Clarify the forecasting objective and the value that a customer obtains from the process.

- Measure the value the customer receives from a leaner forecasting process.

- Identify the flow behind the delivery of value to the customer.

- Determine the pull of the elements for efficiency.

- Establish a continuous improvement process that serves customers best.

So what if you don't have a good forecasting process in place? According to an SAS white paper titled "The Lean Approach to Business Forecasting" [SAS, 2012], the Lean approach to forecasting is motivated by the fact that many existing forecasting process activities are not adding value. The failure can be due to failed systems, flawed forecasting models, or inadequate organization processes, often due to internal politics and management opinions.

A Lean approach consists of gathering data, conducting analysis to add value, communicating the results to management, and streamlining and improving the overall forecasting process.

As the objective is to eliminate non-value-added processes or waste, you need to identify where you are spending resources in the existing process. Figure 6.1 shows a typical, generic forecasting process in goods or service industries.

Demand history and other causal variables are fed into statistical models to create an initial statistical forecast. A forecast analyst may then enter a manual override, based on knowledge of the market and products or information from sources such as sales and customers. Then the forecast, aggregated by family or class of products or

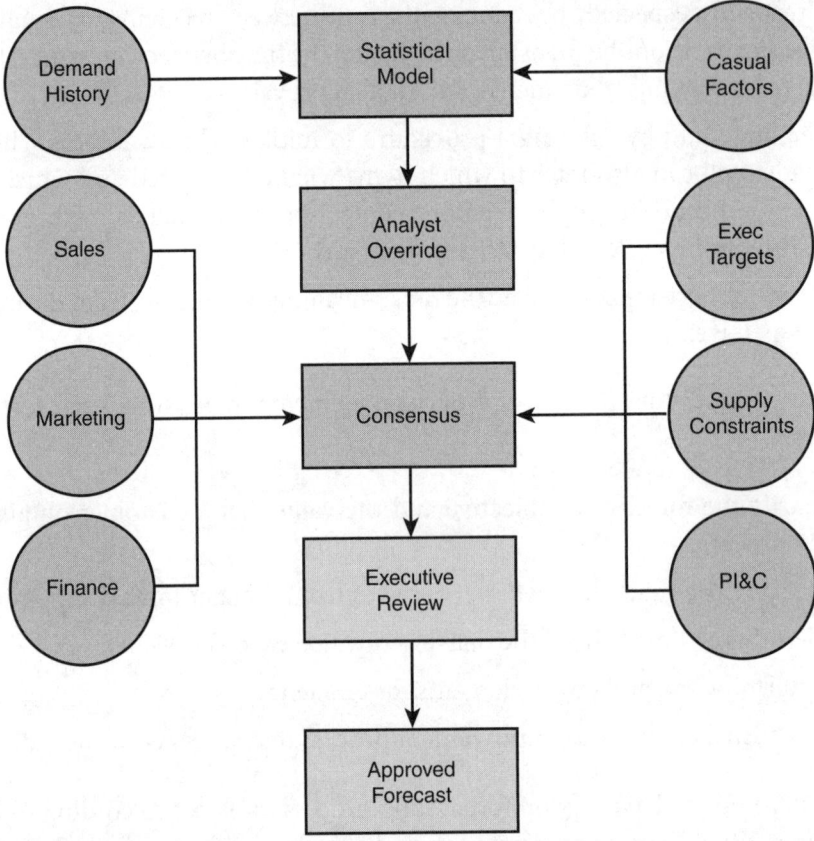

Figure 6.1 Typical Forecasting Process (Goods or Service Industry)

services, is discussed into a consensus, collaborative or sales and operations planning (S&OP) process (see Chapter 8, "Sales and Operations Planning (S&OP)").

As part of the consensus process, members of various functions, such as marketing, sales, finance and accounting, supply chain, and logistics, give their input to the forecasts. This is not limited to the internal supply chain. Many companies have programs with larger customers, such as collaborative planning, forecasting, and replenishment (CPFR), part of which includes collaboration on forecasts with major customers (see Chapter 19, "Collaborative Supply Chain Systems").

A last step in the process, after supply constraints are considered, is the executive approval of the forecast and subsequent supply requirements.

The SAS white paper suggests starting with a "naive" forecast, which takes little effort and basically assumes that actual demand for an item the previous month will be the forecast for the upcoming period. The thinking is that the naive forecast shouldn't be nearly as good as your existing fairly elaborate process. If it is, then something is extremely wrong with your current process.

Whatever software you are using (as discussed later in this chapter), you should gather data on all steps (from Figure 6.1) and participants in your forecasting process. It is human nature to assume that by applying more sophisticated forecasting methods and by developing a more elaborate process and including more management participation in the forecasting process, you will get more accuracy, but this may not be the case, and to determine that you require supporting data and analysis.

SAS suggests using a forecast value-added metric to measure the change in a forecasting performance metric (such as forecast accuracy, positive or negative bias, or mean absolute percentage error [MAPE]) that can be attributed to a particular step or participant in the forecasting process.

The forecast value-added metric allows you to see if each additive change has improved accuracy through added effort (i.e., demand history → statistical model → statistical forecast → management override) by comparing the results of each process activity to the results that would have been achieved without doing the activity.

Forecast accuracy targets shouldn't be unrealistic as perfection is seldom achieved; unpredictable outside factors may impact the forecast. There are a number of other realities of forecasting. For example, product family and aggregated forecasts are more accurate than individual product forecasts, and the further out the time horizon is, the harder it is to forecast demand. It's really more about minimizing variance by setting meaningful targets and putting the appropriate time and number of resources on the development of the forecasts.

As touched on earlier, one fairly common way to do this is using the Pareto principle, or 80/20 rule, which states that a fairly small number of items generate the majority of revenue or profits. These fast-moving products are considered "A" items, and considerable effort should therefore be placed on determining their forecasts, with tighter (i.e., smaller) variance targets. The other items ("B" and especially "C" items, of which there are typically many, usually generating a much smaller amount of revenue or profits) generally have less accurate forecasts and as they don't contribute that much to the bottom line, they deserve less attention in the forecasting process.

Targets may also be tied to other criteria such as to where an item is in its product life cycle (i.e., introduction, growth, maturity, or decline). Typically, new or declining items are more volatile (and rely more on qualitative approaches to forecasting) that growth or mature items, which tend to rely more on statistics for forecasting demand.

Forecasting Technology Options and Requirements

As with most other technology in the supply chain, there are a range of options when it comes to forecasting demand—ranging from simple spreadsheets, to statistical forecasting tools more for analytical use, to "point solutions" that allow for the efficient processing of large numbers of items (see Figure 6.2) and can be integrated with other more comprehensive ERP applications, to one of many integrated modules in a packaged ERP solution.

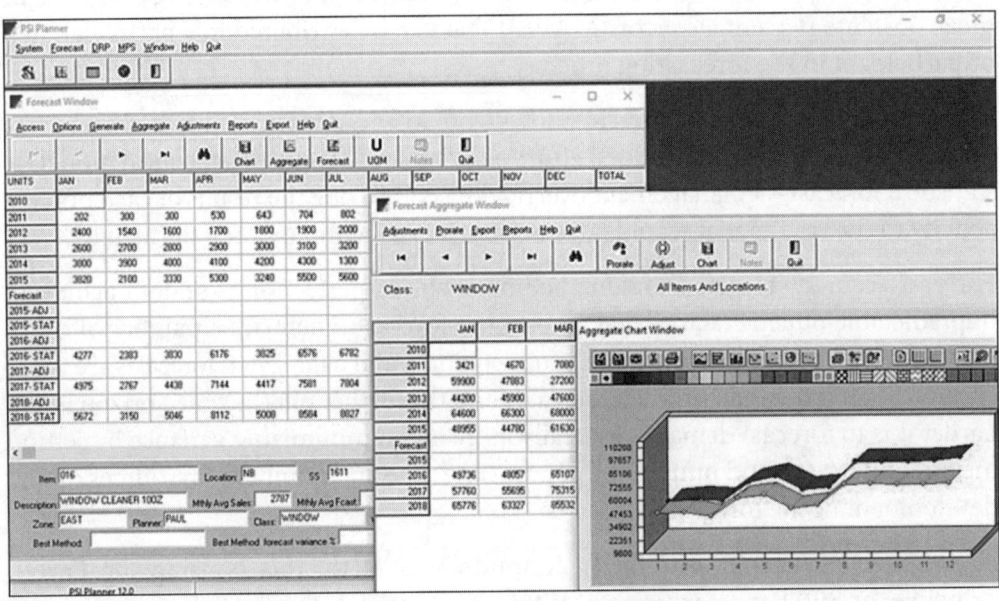

Figure 6.2 Demand Forecasting Item and Aggregate Screen Example (PSI Planner for Windows, Copyright 1998–2016; Printed with Permission from Weeks Software Solutions, LLC)

There are also installed client/server and desktop solutions, as well as Software-as-a-Service (SaaS), on-demand modules found on the web, also known as "cloud software." From a Lean perspective, forecasting software attempts to minimize waste by tracking trends that will affect future demand and accomplish the task by improving the forecasting process in order to eliminate errors or biases in the data and also reducing data latency, which makes real-time (or at least closer to real-time) demand planning possible.

Demand planning tools help an organization reach two core objectives: improved demand forecast accuracy and greater control over demand "shaping" (i.e., influencing of demand to match planned supply).

Demand forecasting systems attempt to accomplish the following:

- **Historical analysis**—Cross-functional or multi-dimensional statistical and judgement-based forecasting processes analyze the historical demand for an individual product at various levels of detail (often factoring in broader macro- and microeconomic trends) to deliver an accurate forecast to supply chain executives.

- **Data separation**—The software gathers a wide range of data to improve past and future demand analysis. Often the data is filtered and separated by product, customer, seasonal, and market information.

- **Demand shaping**—The software factors in data that is related to promotions, advertising, planned introductions of new products, and upcoming competitor activity. It then accumulates and displays detailed plans for future marketing campaigns as well as the projected effects on demand and revenue that may occur.

- **What-if analysis**—The software can perform a series of scenarios and simulations based on what-if factors. The results of these what-if simulations can suggest potential deviations from planned demand, which can then be communicated to other areas of the supply chain to adjust shipments or production.

- **Supply chain communication**—Demand planning systems are typically integrated with other elements of the supply chain management system to drive and change replenishment scheduling as part of the supply or capacity planning process, preventing excess inventory, reducing inventory carrying costs, and also helping to meet targeted customer service levels.

When deciding on specific forecasting software, you need to not only consider your functional requirements but also assess the potential system's functionality and its value to the planning process. You also need an understanding of the efficiency, accuracy, and relevance of the data that is provided, since that data is key to the effectiveness of the system.

As per the forecasting steps previously mentioned, when selecting demand planning software, you should consider the following:

- What specific, detailed data does it collect and interpret?

- How often is the data collected (and how often does the system need to be updated)?

- Will the system be used by itself as more of an analytical tool, does it integrate with supply chain planning, or is it a comprehensive SCM system? Is it compatible with legacy systems? [Harris, 2015]

Lean Forecasting Technology Case Studies

The following sections provide some actual examples of companies that have used demand forecasting technology to reduce waste and improve the efficiency in their forecasting process.

Case 1: Caribou Coffee Automates Forecasting Process with Logility Software

Challenge

Logility, a leading supplier of collaborative solutions to optimize the supply chain, has helped many of its clients to not only implement its forecasting software but gain significant improvements. One such example was Caribou Coffee, with its headquarters and coffee-roasting facility in Minneapolis, and locations across the United States as well as more than a dozen international markets.

Prior to selecting and implementing a forecasting system, like many other companies, Caribou managed its supply chain with spreadsheets. This limited the company's ability to generate accurate and timely estimates of demand to efficiently drive replenishment activities. It needed more accurate visibility to optimize inventory levels while maintaining customer service as product sales expanded.

Approach

The use of Logility's forecasting software enabled Caribou to automate the process of reliably predicting market demand driving efficient sourcing and production, as well as optimal inventory levels throughout the supply chain. By deploying Logility Voyager Solutions, Caribou Coffee's supply chain team went from monthly to weekly demand planning and forecasting. As the company has grown, it has managed to triple its number of SKUs, launched a range of new products, and grown both domestic and international markets.

Results

Through a more automated forecasting process, Caribou Coffee's supply chain team was able to transition demand planning and forecasting from monthly to weekly process while at the same time improving reaction time, inventory turns, and customer service levels, all the while maintaining customer service levels at above 99%. The company was also able to decrease write-offs due to aged inventory and improve inventory turns by 35% [Logility, 2015].

Case 2: Butterball Uses JDA Software to Improve Its Forecasting Process

Challenge

Butterball produces a billion pounds of turkey products every year at five facilities in North Carolina, Arkansas, and Missouri, and ships these products to the 98% of American grocers that carry part of the Butterball line, as well as retailers in more than 30 other countries.

Because Butterball's products are highly seasonal, heavily promoted, and date sensitive, the company faced a number of significant supply chain challenges. To deliver a high level of customer satisfaction, maintain freshness, and minimize obsolete inventory, it needed to have highly accurate forecasts. As a result, Butterball turned to JDA Software (www.jda.com) for help improving and managing its complex short- and long-term forecasting process.

Approach

Butterball's approach had been focused more on the short term, with heavy use of manual data manipulation, which tended to cause wasted time for the planners. JDA demand software allowed the company to focus more on exception monitoring, longer-range planning, and demand shaping.

It has also allowed the company to focus more on product perishability and meeting retailers' different service level expectations and product freshness requirements. Planners now can modify Butterball's forward plans to minimize excess product and maximize customer satisfaction. Planners now have the ability to separate normal demand from promotional demand streams.

Results

Benefits to Butterball include a 28% reduction in obsolescent inventory, a 2% improvement in short-term forecast, and a 50% reduction in long-term forecast bias.

Butterball has leveraged this technology to create a VMI (vendor managed inventory) replenishment model with a key customer (see Part VIII, "Where do we go from here?"). Its increased level of retailer collaboration is providing the building blocks to further expand Butterball's demand network [JDA Software, 2015].

Case 3: Kimberly-Clark Integrates Point of Sale (POS) Information to Improve Its Resupply Process with Retailers

Challenge

Kimberly-Clark makes personal care products, including Kleenex facial tissues, Huggies diapers, and Scott's paper towels, with worldwide sales of $20 billion in 2011.

In 2006, company executives decided to change Kimberly-Clark's supply chain strategy from focusing primarily on supporting manufacturing to meeting the specific needs of its retail and grocery customers.

To do so, Kimberly-Clark realized it would need to include point-of-sale (POS) information about actual consumer purchases to improve the resupply process with retailers.

In 2009, the company used some minimal downstream retail data in its demand-planning software, but for the most part, it relied on historical shipment data for its replenishment forecasts, knowing that forecasts based on historical sales are subject to errors, due to the bullwhip effect; the result was excess safety stock and unsold inventory.

Approach

Kimberly-Clark conducted a pilot program with the software vendor Terra Technology, incorporating POS data into its North American operation. The pilot was successful, and in 2010 Kimberly-Clark licensed and implemented Terra Technology's multi-enterprise demand-sensing solution.

Kimberly-Clark has three retail customers that generate one-third of its consumer products business in North America and provide point-of-sale data that is fed daily into the software to recalibrate the shipment forecast for each of those retailers. The software evaluates any new data inputs from the retailers, along with open orders and the legacy demand-planning forecast, to generate a new shipment forecast for the next four weeks. Kimberly-Clark also uses that forecast to guide internal deployment decisions and tactical planning.

The software processes data from the retailers, such as point-of-sale information, inventory in the distribution channel, shipments from warehouses, and the retailer's own forecast and reconciles that data to create a daily operational forecast. It also identifies patterns in the historical data to determine how much influence each input has on the forecast. One example might be that POS is found to be the best predictor of a shipment forecast on a three-week horizon, but actual orders and legacy demand forecasts could be the best predictor for the current week.

Results

By incorporating demand signals from key retail customers into its shipment forecasting process, Kimberly-Clark has realized substantial improvements, such as being able to develop a more granular metric for forecast errors. Ultimately, it found a reduction in forecast errors of as much as 35% for a one-week planning horizon and 20% for a two-week horizon.

Furthermore, forecast accuracy improvements, which resulted in reductions in safety stock, have helped Kimberly-Clark reduce its overall inventory and reducing its finished-goods inventory by 19% over the previous year and a half [Cooke, 2013].

While an improved forecasting process that uses available technology is critical to all businesses, the typical next step for goods and some service organizations is to answer the questions of "how much" and "when" to produce or purchase additional inventory. Chapter 7, "Master Production Scheduling (MPS)," discusses this.

7

Master Production Scheduling

Master Production Schedule (MPS) Defined

A master production schedule (MPS) takes a business plan and other inputs from financial plans, customer demand, engineering, and supplier performance to create a comprehensive product manufacturing schedule for independent demand inventory (i.e., end items or finished goods). The MPS covers what is to be assembled or made, at what time, and with what materials, as well as the cash required during each week of a relatively short-range planning horizon. MPS is a key driver of material requirements planning (MRP), which determines raw material and component requirements, known as *dependent demand* inventory (see Chapter 9, "Material Requirements Planning [MRP]") as well as a short-term manufacturing schedule (see Chapter 13, "Short-Term Scheduling").

The MPS must be copacetic with the aggregate production plan (see Chapter 8, "Sales and Operations Planning [S&OP]"), which attempts to create a supply plan that satisfies demand at the lowest cost. As the process moves from planning to execution, each step must be tested for feasibility in terms of staffing, machine, and material constraints.

Rough cut capacity planning (RCCP) involves quick checks on a few key resources to implement the MPS in order to ensure that it is feasible from the capacity point of view. The MPS and RCCP are developed interactively. RCCP is used to determine the impact of the MPS on the key or aggregate resources, such as human or machine hours. Rough cut capacity plans can be "finite," or constrained, because they have to operate within certain constraints, or they can be "infinite," or unconstrained, leaving adjustment decisions to the expertise and knowledge of the planner.

Inputs for a master production schedule may include forecasted demand, production costs, inventory, customer needs, lot size, production lead time, and capacity. Inputs may be automatically generated by an ERP system.

A typical output for a finalized MPS is a production plan, in a format often referred to as a PSI (production, sales, and inventory) report (see Figure 7.1). A PSI report may include quantities to be produced, staffing levels, quantity available to promise, and projected available balance. It is typically generated at the item level for a particular sourcing facility (internal or outsourced) or market zone and shown in weekly or monthly time planning buckets.

ABC Company
PSI Report

December 4,2015

Plant/Market Zone : EAST
Item : 026
Description : GLASS CLEANER
Safety Time : 1.5
Safety Stock: 3700

Period	Past Due	04Dec15	07Dec15	14Dec15	21Dec15	28Dec15	04Jan16	11Jan16	18Jan16	25Jan16	01Feb16	08Feb16	15Feb16	Total
Periods of Supply	0.86	25.65	10.42	9.42	9.26	7.79	6.52	5.52	4.42	3.66	2.63	2.60	2.61	
Beginning Inventory	7604	23705	23781	21499	22239	19837	17209	14637	11905	9273	7161	7130	7115	256998
Forecast	0	204	1022	1022	1022	988	852	852	852	852	1043	1043	1043	10795
Customer Orders	7502	0	0	232	0	0	0	0	0	0	0	0	0	7734
Dependent Demand	1140	720	1260	1260	1380	1560	1800	1800	1860	1680	1680	1620	1680	19440
Scheduled Receipts	24943	1000	0	3022	0	0	0	0	0	0	0	0	0	28965
Planned Orders	0	0	0	0	0	0	0	0	0	420	2700	2640	2760	8520

Plant/Market Zone : WEST
Item : 026
Description : WINDOW CLEANER
Safety Time : 3
Safety Stock: 1100

Period	Past Due	04Dec15	07Dec15	14Dec15	21Dec15	28Dec15	04Jan16	11Jan16	18Jan16	25Jan16	01Feb16	08Feb16	15Feb16	Total
Periods of Supply	0.0	0.98	0.58	0.57	0.55	0.56	0.49	0.46	0.43	0.40	0.45	0.46	0.46	
Beginning Inventory	0	60	179	173	167	221	367	344	321	358	455	462	469	3576
Forecast	0	61	306	306	306	394	743	743	743	743	1013	1013	1013	7384
Customer Orders	0	0	0	0	0	0	0	0	0	0	0	0	0	0
Dependent Demand	0	0	0	0	0	0	0	0	0	0	0	0	0	0
Scheduled Receipts	0	0	0	0	0	0	0	0	0	0	0	0	0	0
Planned Orders	60	180	300	300	360	540	720	720	780	840	1020	1020	1020	7860

Plant/Market Zone : ALL
Item : 026
Description : WINDOW CLEANER
Safety Time : 2.5
Safety Stock : 6300

Period	Past Due	04Dec15	07Dec15	14Dec15	21Dec15	28Dec15	04Jan16	11Jan16	18Jan16	25Jan16	01Feb16	08Feb16	15Feb16	Total
Beginning Inventory	7673	26190	27730	25435	28944	26641	24263	21598	18917	16244	13991	14012	13973	265419
Forecast	0	460	2303	2303	2303	2378	2673	2673	2673	2673	2679	2679	2679	20476
Customer Orders	7502	0	0	232	0	0	0	0	0	0	0	0	0	7734
Dependent Demand	1140	720	1260	1260	1380	1560	1800	1800	1860	1680	1680	1620	1680	19440
Scheduled Receipts	24943	1000	0	3022	0	0	0	0	0	0	0	0	0	28965
Planned Orders	180	720	1260	1260	1380	1560	1800	1800	1860	2100	4380	4260	4440	27000

Figure 7.1 Production, Sales, and Inventory (PSI) Report Example

The technology used to generate a master production schedule can range from fairly complex spreadsheets to modules within ERP or supply chain planning systems. Integrated solutions offer the benefit of being connected to the aggregate, material requirements planning, and short-term scheduling systems, creating more efficient and effective results.

Lean Scheduling

Scheduling is very important to manufacturing, as it focuses on the allocation of scarce resources to tasks over time. Solving scheduling problems can be very complicated, and as a result, it's not always possible to find the best possible solution in a reasonable time frame.

Heuristic methods, rules or methods that come from experience (i.e., shortcuts), have been developed in order to find near-optimal solutions in comparatively short periods of time. However, heuristics often applied in practice dispatch rules that have minimal computational complexity and are simple to implement.

It is difficult to execute a schedule precisely, but a main objective is to be able to accommodate and anticipate uncertainties before they occur or have a plan to counteract them.

Production planning in a Lean environment requires leveling out the production schedule by smoothing out the peaks and valleys. In order to deliver this kind of Lean schedule, an organization must be able to make quick changeovers from one product to another and must be able to produce in small lot sizes. The best way to do this is with a demand "pull" type of system as opposed to the traditional "push" or large lot production.

In a Lean operation, products are often produced to a buffer called a finished goods "supermarket" rather than directly to customer orders, as that might not be feasible due to the relatively small quantities on each order.

The concept of takt time is critical to shifting to level scheduling. *Takt time* can be thought as the "heartbeat" of the plan: It is the rate at which a finished product (usually a family of products) needs to be completed in order to meet customer demand.

Once the takt time is known, the process bottleneck, a resource that requires the longest time in operations of the supply chain for certain demand and therefore limits capacity or throughput, can be determined. This "pacemaker" then drives the pull process both upstream and downstream.

Various Lean methods can be used to relieve bottlenecks (and level production) when they stop the process from meeting the required takt time:

> Levelling of production by both volume and product mix. Products are not manufactured in accordance with the sequence of customer orders; rather Heijunka [a Japanese word that means "leveling"] calculates the total volume of orders in a period and levels them out so the same amount and mix are manufactured each shift/day. Small-scale Lean organizations use spreadsheets to schedule their production in order to create Heijunka.
>
> However, IT systems are a crucial addition for most organizations and yield significant benefits. With the addition of the Internet, this has exploded its potential. The spreadsheet approach can be helpful in a pilot context or small-scale

entities, but it is questionable whether spreadsheets are a scalable technology in larger organizations, as it's known that they encounter issues with data reliability when used in isolation. Hence, there is a requirement to enable the scheduling method to integrate with an organization's Enterprise Resources Planning (ERP) or Supply Chain Management (SCM) systems. [Salman et al., 2010]

Furthermore, executing a Lean JIT type of schedule requires partnerships with suppliers, with future production schedules that drive purchasing requirements for parts and supplies shared through visual or electronic kanbans (systems to control the supply chain from a production viewpoint). In many cases, suppliers replenish automatically through a vendor-managed inventory (VMI) system, as discussed in Chapter 19, "Collaborative Supply Chain Systems."

MPS/Production Planning Technology Options and Requirements

It is still somewhat common in many organizations for a production plan to be generated using a spreadsheet. Unlike with other supply chain applications, such as forecasting and MRP software, there isn't really much in the way of standalone MPS software (usually including some kind of rough cut capacity planning capability). However, MPS can usually be found as a module in many ERP systems as it needs to be tightly integrated with information contained in these types of systems, such as forecasting, inventory, purchasing, and manufacturing scheduling, as well as aggregate planning and manufacturing short-term scheduling systems and/or modules. Examples include the JDA Master Planning module (www.jda.com) and Logility Voyager Manufacturing Planning (www.logility.com).

Using technology such as this for production planning can help create reliable, feasible master plans that can drive manufacturing and cost-efficiencies while improving inventory management. These tools usually featuring time-phased views of each product and manufacturing resource load. They do this by generating constraint-based schedules and capacity plans that maximize throughput (a major benefit or at least goal of most Lean processes).

Lean and ERP for Production Planning: A Fine Balancing Act

A corporate-wide enterprise resource planning (ERP) system can challenging when it comes to working with Lean initiatives on the plant floor as there can be some major conflicts. One conflict that often arises is between materials planning and production scheduling.

ERP tends to use a top-down "push" approach, depending heavily on sales forecasts for materials planning, while Lean is based on a customer demand "pull" production schedule, keeping minimum inventory and using a kanban system that replenishes materials and parts only when needed. For example, TRW's European Foundation Brakes Division, which manufactures brake calipers, drums, boosters, antilock braking systems, and electronic stability-control systems, as well as various suspension components, has eight plants in five countries and has successfully used a strategy of keeping ERP mostly outside Lean-driven plants.

Internally, the plants in TRW's European Brakes Division are focused on their Lean initiative. However, when dealing with customers and suppliers, ERP provides them with transactional information. On a weekly basis, a logistics planner in each plant, with the help of an automated calculation, takes customer order data from the ERP and levels demand and produces a similar number of parts of every product every day for the upcoming weeks.

ERP provides the customer order information, but each plant builds its own level calculation of demand from that data. As they are building parts at the same pace that customers need those parts, sequencing of orders is generally not a problem.

Finally, ERP is used again at the end of the production process, when brake parts are finished and ready to be shipped to customers [Bartholomew, 2012].

Lean Production Planning and Technology Case Studies

The following sections provide some actual examples of companies that have used technology to reduce waste and improve the efficiency in their production planning process.

Case 1: Energy Bar Company Upgrades Production Planning Technology

Challenge

A fledgling energy bar company wanted to create a flexible production plan driven by various demand projections over a 12-month period. It wanted to use a low-cost strategy to schedule hiring of additional workforce, ongoing raw material purchases, capital equipment purchases, and production facility decisions for current and forecasted demand. It initially did calculations in a spreadsheet, manually adjusting production and purchasing variables. The company soon realized that changes in demand caused issues in other areas that were not easily accounted for in the spreadsheet, causing this form of production planning to be a trial-and-error exercise.

Approach

ORM Technologies (www.orm-tech.com), which has created the Optimization and Operations Research Engineering software system, implemented its Production Planning software module with the energy bar client to include all known production, cost, and staffing constraints to produce a 12-month plan based on current and forecasted product demand.

The new technology allowed the energy bar client to do the following:

- Quickly create production plans that maximize profit with dozens of different demand profiles
- Select the optimal timing for hiring additional workforce, purchasing raw material, and making facility decisions to meet current and forecasted demand
- Perform what-if analysis with current and forecasted product demand as well as facility, staff, production, raw material, and packaging costs and capacities

Results

The system output provided the company with the information needed to make staffing and facility decisions as well as capital equipment and reoccurring raw material purchases. It also allowed the company to save much-needed capital by delaying a move to a new facility by several months and optimizing the staffing plan. The client can now run what-if analysis whenever it likes to determine the production, cost, inventory, and staffing effects of changes in demand to plan ahead rather than operate in a reactive mode [ORM Technologies, 2015].

Case 2: Global Raw Materials Supplier Integrates MPS Process to Minimize Waste

Challenge

A global manufacturer was running a production and material planning process that needs to be highly integrated with demand (customer orders) in order to minimize waste on the manufacturing floor while allocating demand across all global plants to optimize output. It used a production planning data mart to provide synchronization and a consolidated production view across each of the manufacturing facilities. This data mart urgently needed to be updated, both in terms of the technology and the logic driving the process.

Approach

Corporate Technologies (CTI), a systems integrator and solutions provider (www.cptech.com) was brought in to reverse engineer the business logic and implement a new production planning data mart to reduce the cost of ownership while improving the maintainability of the system.

The strategic plan for all business units was to use an integrated SAP architecture, applying SAP Business Objects Data Services as the data integration technology across the manufacturing systems. This involved analyzing hundreds of different data flows from thousands of different data sources. Business rule logic was reengineered in SAP Business Objects Data Services and tested for business integrity.

Results

Ultimately, the updated technology and logic significantly reduced IT maintenance costs and improved data inputs to the manufacturing planning process [Corporate Technologies, 2015].

Case 3: Nestlé Innovates Its Milk Production Planning
Challenge

Nestlé Pakistan operates the largest milk collection operation in Pakistan, working with approximately 190,000 farmers in the provinces of Punjab and Sindh. The company produces a variety of dairy products, including milk, powdered milk, cream, tea whiteners, and yogurt.

Even though demand for milk products is fairly constant year round, not surprisingly, milk production varies significantly from season to season. Because the content of milk (e.g., fat) from farmers varies, Nestlé has the added problem of managing its production capacity in the most efficient way with both supply constraints and demand variations.

Although Nestlé had invested in an ERP system, it was unable to automate the complex milk production planning process, and the manual planning that was being used had its own set of limitations.

To manage content variation, Nestlé prepares a production plan every month, using a number of estimates and assumptions. Its typical production plan is generated in a spreadsheet by a production planner, following a set of objectives with a large number of rules and constraints. The usual production plan includes the following objectives:

- Minimize milk waste.
- Reach planned production quantities for each SKU.
- Produce material ahead for future months, when fresh milk supply may be inadequate.
- Efficiently utilize imported material.
- Create requests in advance of anticipated material shortage.
- Utilize available plant capacity for bulk production and line capacity for packaging.

The following rules and constraints are used in planning:

- Maintain production ratios between products when increasing or decreasing production.
- Follow plant and packaging line capacities and maintenance schedules.
- Track raw material availability and stock expiration dates.
- Select the most appropriate bill of material (BOM) from a large of number of potential recipes, which depends on raw material and fresh milk availability.

Nestlé's SAP ERP system did not fully capture the complexity of the milk production planning problem and thus the company relied on manual planning, which has multiple flaws:

- **Suboptimal planning**—The production planning process produced suboptimal plans, largely as a result of trying to solve a complicated multi-constraint optimization problem primarily using instinct, approximations, and heuristics from past experience. This manifested in line capacity remaining idle, using less desirable BOMs, etc.
- **Time-consuming and repetitive**—The entire planning process, which included the generation of multiple plans, took a considerable amount of time. The process had to be repeated each month and took a significant portion of the month to complete.
- **Error prone**—The production planner had to consider a wide range of conditions, rules, and constraints when creating the monthly production plan. As a result, human errors were fairly common.

- **Expert dependent**—The production plan at Nestlé is dependent on a very small number of experts. Training new resources is difficult and takes considerable time.

As a result, Nestlé decided to develop an automated solution in order to simulate multiple what-if scenarios using rules and constraints to prepare a final production plan that met all set parameters.

Approach

Nestlé engaged Techlogix (www.techlogix.com), an IT services, consulting, and business solutions company, to create a customized planning system. Techlogix proposed a multi-phased approach, starting with a feasibility study and high-level solution design; followed by a detailed design, build, and delivery of the solution; with a final phase of maintenance and ongoing support.

Techlogix designed and delivered a web application to be used as a simulation engine to produce a production plan along with a variety of reports. The reports helped the planner look at different aspects of the production plan.

The production planning algorithm has two distinct aspects: planning and scheduling. The planning algorithm uses fresh milk quantity, raw material stock quantity, and other inputs to determine the optimal production of bulk material. This quantity is the input to the scheduling algorithm, which attempts to schedule production on plants within specified capacities. If required production capacity is not available, the planning algorithm adjusts production and attempts a reschedule to find the best fit. The reports that are generated allow a production planner to view production in terms of bulk produced, SKUs produced, plant and line capacities used, and so on.

The goal during testing was to manage multiple production lines based on the system output. Five months of parallel runs were conducted, during which time the system output was compared with the manual plan and checked against all constraints. Toward the end of testing, the system was able to consistently generate plans that met all criteria and significantly improved upon the manual plan.

Results

The automation of the production planning process now takes only a few hours compared to the days of effort previously required, and there is now more time for maneuverability. The new production planning output is more precise and accurate than before, which helps reduce costs and eliminate waste [Techlogix, 2015].

Case 4: Durabuilt Windows and Doors Balances Lean and ERP for Production Planning

Challenge

Durabuilt, a manufacturer of windows and doors, operates a 180,000-square-foot plant with 450 employees. It began using the ERP system Cantor in 2008 and faced the challenge of integrating Lean practices at its plant over a span of three years, which was a real balancing act as the company needed to survive and prosper.

Approach

In the past, Durabuilt's workflow on the production floor ran through ERP, and the company would produce 50 or 100 boxes before going to the next order. As a result, large quantities of boxes waited for hours to be used. Before it began using ERP, Durabuilt sent paper documents to purchasing to order materials and then to receiving to wait for the materials to come in.

Results

Durabuilt's IT staff had to adapt the ERP system to support the Lean management principles used at the facility. As part of the Lean initiative, the plant had modified the assembly line to single-piece flow and needed to configure the ERP system to support that process. Now, instead of having batches of parts sitting around in boxes waiting for hours to be used, the parts are brought to the line as needed.

Management strongly believes that without an ERP system, Durabuilt's Lean initiatives wouldn't work. For example, the company no longer has a transfer of paperwork for purchasing as orders for materials go straight to suppliers. The process is now faster, and Durabuilt is not missing anything due to human error. Material lead time is built right into the system, providing an accurate delivery date to the customer up front.

In the new process, salespeople enter orders into the ERP system, and those orders then go to the scheduling department. Schedulers then create and adjust the production schedule in ERP and fine-tune it daily, factoring in variables such as capacity of individual manufacturing lines and complexity and time sensitivity of customer orders. Schedulers then put consideration into the transport schedule for Durabuilt's fleet of trucks that deliver windows and doors to customers [Bartholomew, 2012].

Some companies start with MPS/production planning and then aggregate those plans for collaboration purposes; others start at the aggregate and then disaggregate for production planning. In either case, they are directly linked and must be in sync. Chapter 8, "Sales and Operations Planning (S&OP)," discusses aggregate or S&OP, its impact on a Lean supply chain, and how technology can greatly assist in this process.

8

Sales and Operations Planning (S&OP)

Sales and Operations Planning (S&OP) Defined

In its simplest terms, *sales and operations planning (S&OP)* is a process a business uses to ensure that supply can match demand, at least on the aggregate. (S&OP is therefore also often referred to as *aggregate planning*.) With S&OP, executive-level management regularly meets and reviews projections for demand, supply, and the resulting financial impact (typically integrated with the results of more detailed work, described earlier, when developing forecasts and in some cases production plans). S&OP is a decision-making process which ensures that tactical plans in every business area coincide with the company's business plan. The net result of the S&OP process is that a single operating or aggregate plan is created that allocates company resources.

An even broader definition of S&OP has emerged in recent years, offering an even greater impact from a Lean perspective on a supply chain: S&OP can be considered integrated (or advanced) sales and operations planning (also referred to as *integrated business planning [IBP]*), which represents the transition of S&OP from its production planning origins into the fully integrated business management and integrated strategy and financial planning process it is today. There is now more effort put on not only looking at a business's internal supply chain but also the extended supply chain, both downstream, toward customers, and upstream, toward suppliers. This effort has been enhanced by many of the collaborative programs and technology-enabling tools used in demand and supply chain planning, such as vendor-managed inventory (VMI) and Collaborative Planning, Forecasting, and Replenishment (CPFR), discussed in Chapter 19, "Collaborative Supply Chain Systems."

S&OP increases teamwork between departments and helps align the operational plan with the strategic plan. It is a process in which various targets are set (i.e., forecast accuracy, inventory turns) and progress against the strategic and operational plans are reviewed in a series of monthly meetings.

The objective of S&OP is to come to consensus on a single operating plan that meets forecasted demand while minimizing cost over the planning period. It should allocate people, capacity, materials, and time at the least possible cost, while ensuring the highest customer service possible.

The executive S&OP process (see Figure 8.1) actually sits on top of the number-crunching and analysis being done at a lower level of the organization (e.g., item forecast generation, production planning). It involves a series of monthly meetings prior to a final S&OP executive-level monthly meeting that are used to create, validate, and adjust detailed demand and supply plans.

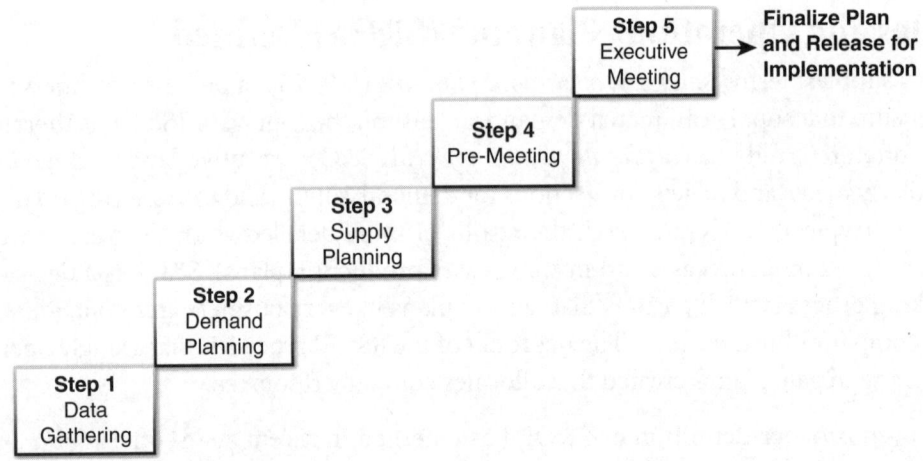

Figure 8.1 Steps in the S&OP Process

The following S&OP meetings occur before the executive-level meeting:

- **Demand planning cross-functional meeting** (step 2)—Generated forecasts are reviewed with a team that may include representatives from supply chain, operations, sales, marketing, and finance. Prior to this meeting, forecasts have already been generated statistically and aggregated in a format that everyone can understand and confirm (e.g., sales might want to see forecasts and history by customer in sales dollars).

- **Supply planning cross-functional meeting** (step 3)—This meeting occurs after confirmed forecasts have been "netted" against current on-hand inventory levels to create production/purchasing plans. These data are usually reviewed in the aggregate by product family in units, for example (see Figure 8.2).

First Quarter			Second Quarter			Third Quarter		
January	February	March	April	May	June	July	August	September
50,000	30,000	55,000	60,000	80,000	150,000	150,000	125,000	80,000

Aggregate Plan

Months:	January				February			
Aggregate Plan Quantity:	50,000				30,000			
Weeks:	1	2	3	4	5	6	7	8
MPS Quantity:								
26" Boys Blue	10,000		10,000		5,000		5,000	
12" Boys Red		12,500		12,500	8,500		8,500	
12" Boys Yellow		5,000					3,000	

Master Production Schedule (MPS)

Figure 8.2 Aggregate Plan Versus Master Production Schedule

- **Pre-S&OP meeting** (step 4)—During this meeting, data from the first demand and supply meetings are reviewed by department heads to ensure that consensus has been reached.

The discussions from this series of monthly management meetings highlights issues and looks at possible resolutions before the outcome of the discussions is presented to the senior management team as a series of issues to be resolved. These issues form the basis of the monthly executive S&OP meeting (step 5).

S&OP and Lean

Lean teams plan and execute on a shop floor level, but S&OP can be a great tool for making the connection between Lean kaizen event goals and objectives and corporate ones. Remember that inventory is one of the eight wastes and covers variability in a process. Through the use of S&OP, inventory and other supply chain operating costs can be directly planned and controlled. In general, there are two ways to reduce inventory: more accurate forecasts and shorter cycle times. The S&OP process attempts to improve and control both of these.

From a Lean perspective, a robust S&OP process acts as both a planning *and* control method at an executive management level. Various metrics indicating the level of waste (e.g., forecast accuracy, inventory turns, on-time and complete shipments) are benchmarked externally to set objectives—as well as match the company's strategic plan—and are measured to determine when things are in or out of control.

According to an Aberdeen Group study in 2010, four key performance criteria distinguish best-in-class S&OP:

- Forecast accuracy
- Perfect orders delivered complete and on time
- Cash to cash cycle
- Gross profit margin [Viswanthan, 2010]

All of these, but especially the first three, are measurements of how lean a company is in that the lower the score, the more "variability" the system has, which leads to many of the wastes we have discussed.

Working Together

In their 2006 book *Sales and Operations Planning—Best Practices*, John Dougherty and Christopher Gray point out that in a study of 13 best practice companies, Lean and S&OP are complementary. According to one of their best-practice clients, "continuous improvement is embedded in the S&OP process, and continuous improvement cannot be maximized without S&OP" [Dougherty, 2006]. They go on to point out:

> If you create a manufacturing environment where material flows with minimum waste (Lean), but you can't predict capacity and material availability problems in enough time to avoid them (S&OP), you will inevitably revert to firefighting, finger-pointing and poor results. Similarly, if you do an excellent job of future planning but have poor flows, you can almost count on higher inventory levels, longer lead times, and lower profitability. … Traditionally Lean manufacturing has been stronger on workplace management; S&OP on decision-making for the future. The tools and methods of Lean manufacturing have tended to look most closely at the plant, and its immediate customers and suppliers, mostly over a short horizon. This leads to improvements like: "shorter, quicker, fewer, lower cost, more flexible, and better aligned." S&OP provides distance vision—providing the ability to predict capacity and material availability problems before they become crises, to identify market issues while they are still opportunities, and to prioritize improvements in a way that will create the most favorable results. [Dougherty, 2006]

Tomorrow's supply chain will be driven, among other things, by effective S&OP, which allows for effective supply chain planning; balances new and current products and services; employs timely, effective replenishment; enables timely success/failure measurement; and, with the help of technology, can generate data analysis and correction.

When S&OP is implemented in a company that is focused on a team-based continuous improvement process, from top to bottom, success can be ensured.

S&OP/Aggregate Planning Technology

Many modules of ERP systems now offer a variety of dashboards that provide the ability to drill down into details about demand and supply information. However, it is still not uncommon for Excel templates such as the one shown in Figure 8.3 from Weeks Software Solutions, LLC (www.ezmrp.com) to be used to integrate with anything from other production planning spreadsheets to modules in ERP systems.

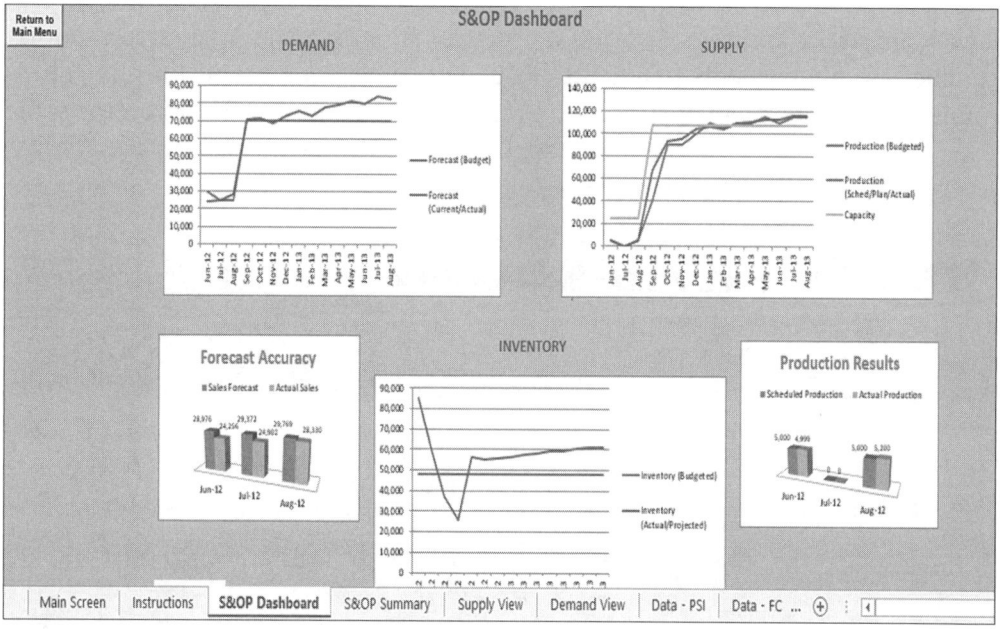

Figure 8.3 S&OP Excel Spreadsheet Template—Dashboard Screen (PSI Planner for Windows™, Copyright 1998–2016; printed with permission from Weeks Software Solutions, LLC)

At the other end of the spectrum are companies such as SAP (www.sap.com), which is powered by the HANA in-memory data platform, which can be deployed as an on-premises installation or in the cloud. It is best suited for performing real-time analytics (see Chapter 18, "Measurements, Metrics, and Analytics") and developing and deploying real-time applications. Using HANA with S&OP offers a unified model of demand, supply chain, and financial data—at any level of granularity and dimension in real time.

The linkages from the dashboard are driven from demand and supply planning activities that have been aggregated into product classes or families. It is therefore not surprising that they are typically integrated on the aggregate with forecasting applications on the demand side. On the supply side, they can be fed data from MPS, in many cases, from constrained/unconstrained, or "rough cut," production (or capacity) planning (RCCP) modules in ERP or supply chain planning systems such as those found in Logility (www.logility.com) and JDA (www.jda.com) software. These type of systems or modules allow for optimization of staffing, material, and machine constraints of facilities or even production lines to minimize resource constraints—sometimes reducing manufacturing planning time as much as 75%.

The example in Figure 8.4 shows a scenario in which the aggregate monthly production requirements (using established crewing, equipment run rates, minimum lot sizes, changeover times, etc.) for a class of products all run on the same

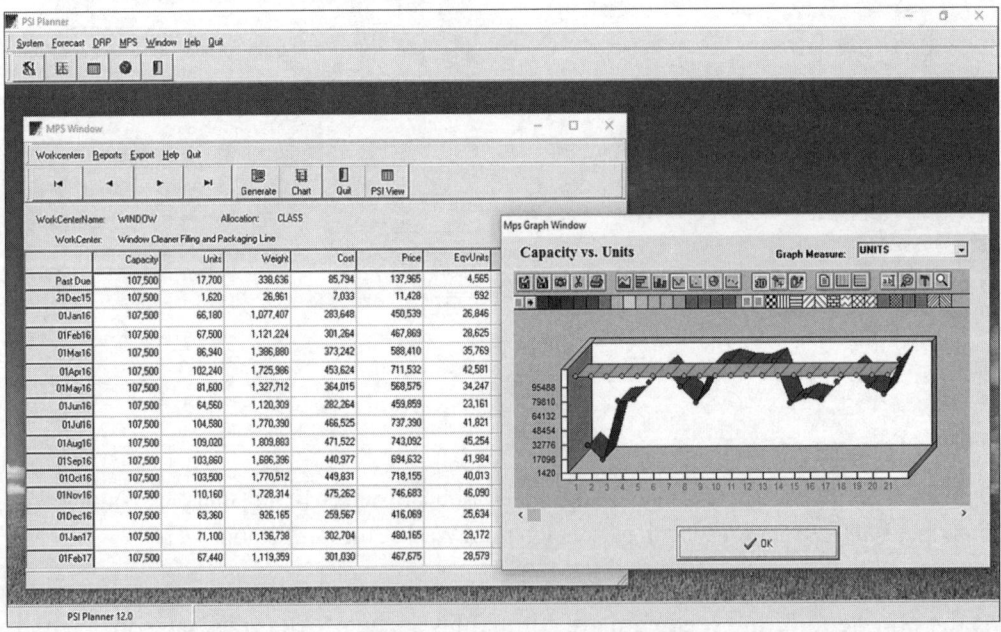

Figure 8.4 Unconstrained Production Plan Screen (PSI Planner for Windows™, Copyright 1998–2016; printed with permission from Weeks Software Solutions, LLC)

filling and packaging line are beyond the stated monthly capacity for some of the planning months. In the case of unconstrained capacity planning software, the planner would use his or her own knowledge, judgment, and other information to handle the situation. In the case of constrained capacity planning software, the software itself, using rules (e.g., priorities, sequencing) established by the planner, automatically creates an optimal solution for the scenario.

It is important to consider a range of options, including demand options such as cutting, delaying, and backordering customer orders, and supply options, such as running overtime and hiring temporary workers (short term), adding shifts and using contract packagers (medium term), and adding new production lines and even facilities (long term).

Lean S&OP/Aggregate Planning and Technology Case Studies

The following sections provide some actual examples of companies that have used technology to help reduce waste and improve the efficiency of their aggregate planning processes.

Case 1: Infineon Technologies AG Takes Planning to the Next Level with JDA S&OP

Challenge

Infineon Technologies AG provides semiconductor and system solutions focused on energy efficiency, mobility, and security. The company has customers worldwide, including auto manufacturers, industrial electronics companies, chip card and security businesses, and information and communications technology companies. Infineon's industry is known for involving volatile demand, long product lead times, significant capital investments, and complex product and supply networks. It is therefore critical for Infineon to be able to identify and respond to demand changes while balancing global production capacity across its more than 20 facilities.

Approach

Infineon made the decision to reengineer its planning processes and tools so it could respond quickly to market changes across all planning levels and areas of the company. The company determined that it needed new technology to accomplish this goal. The reengineering took two years and was jointly managed by the information technology and business departments.

When Infineon decided to reengineer, it was using homegrown sales and operations planning tools that didn't work due to limitations in scalability and integration. Infineon chose JDA Software for this project as the two companies had worked together in the planning area in the past.

The JDA consultants tried to understand their processes first and then tried to match them with the appropriate JDA solutions. Through this partnership, JDA was able to develop an interactive rough-cut capacity planning functionality within JDA S&OP. The new tool enables Infineon to synchronize demand, supply, capacity, and load planning into one multidimensional view.

Results

Sales forecasts, pricing, and production capacity are now reviewed and adjusted in real time in one simulation model that gives Infineon an overview of its operational and financial plans. The company can then quickly react to changes in the market. It can also shape demand to meet production constraints or identify a need for additional capacity (e.g., at a subcontractor site).

As a result of implementing the S&OP solution, Infineon has done the following:

- Reduced its planning effort by more than 30%.
- Cut the lead time for its rolling forecast from four weeks to two weeks.
- Decreased planning errors up to 90%.
- Reduced its "churn" (i.e., making minor adjustments to the plan) by a factor of 10.
- Improved forecast accuracy.

Most importantly, Infineon now has a new collaborative planning approach that enables the business to be more agile and responsive so that demand, supply, capacity, and load are always synchronized and calculated together [JDA, 2014].

Case 2: Accelerated S&OP Collaboration at Lance

Challenge

Lance produces and distributes snack foods, including cookies, crackers, nuts, and potato chips, largely under the Lance, Cape Cod, and Tom's brand names. Lance products are distributed via direct store delivery, with more than 1,400 sales routes and relying on independent distributors and direct shipments to retail customer locations. The company's challenge was to increase visibility, improve forecast accuracy, and increase shelf freshness for consumers.

Approach

To reach its goals, Lance implemented Logility Voyager Sales and Operations Planning software, which transforms information from sales, production, finance, marketing, transportation, and procurement into one central database. It allows S&OP to work from a one-number system, saving time and reaching better clarity. With this S&OP solution in place, Lance determined that it could cut hours and days from its planning process, streamline the planning cycle, and complete analysis in a fraction of the time.

The goal was to allow Lance to have greater visibility into its business so it could identify its most profitable customers and channels, optimize its product mix, improve procurement strategies, and maximize margins. Logility's S&OP best practices allows Lance to compare multiple what-if scenarios, evaluate critical decisions, and prepare contingency strategies to mitigate risk.

Results

Since implementing the solution, Lance had improved forecast accuracy, reduced inventory days on hand, decreased finished goods storage, and improved S&OP collaboration. The new technology has allowed Lance to do the following:

- Improve forecast accuracy from 50% to 70%.
- Reduce inventory days on hand by 20%, from five days to four.
- Decrease finished good storage by two warehouses.
- Accelerate S&OP collaboration.
- Improve acquisition integration effectiveness [Lance, 2015].

Case 3: Continental Mills Increases Productivity with S&OP Process

Challenge

Continental Mills has been manufacturing products such as pancake mix, baking mixes, drink mixes, and breading and batters since 1932. It has three manufacturing facilities located throughout the United States. The current supply chain system and processes have been struggling to successfully cope with the increasing complexity of business today, including many new SKUs and customers.

Continental Mills's S&OP process was supported from the top down, with senior managers participating on a monthly basis. Up to that point, the supply chain team had calendar discipline but with a very labor-intensive process.

The team faced many challenges, including having many versions of spreadsheets emailed back and forth, some with corrupted data. No one knew how much time and effort they put into chasing numbers and version control, but it was substantial, and in many cases they found out later on that none of it had been necessary.

In addition, separate divisions looked outwardly in different ways, each division having forecasts at a different levels of detail.

Approach

As a result of its difficulties, Continental Mills looked for a new solution to manage that growth, and it ended up selecting Logility's Voyager Solutions.

One benefit of the Logility solution is that it provides Continental Mills with a flexible solution that allows each of the four divisions to manage its business in its own way and in different levels of detail but still gives them all a comprehensive corporate view through a single hierarchical structure.

Continental Mills has used the Voyager Demand Planning system to convert its businesses from its hard-to-maintain forecasting process using spreadsheets to a statistical forecast that reduced the time and effort previously dedicated to administration and crunching numbers. The company can now do a much better job of driving the forecast and analyzing options, with a separate view for each of the businesses. It has saved an estimated 40 to 50 hours of spreadsheet manipulation from the monthly S&OP process.

The S&OP software has allowed Continental Mills to streamline the production planning process and improve capacity planning for all its manufacturing facilities.

Results

The new system allowed Continental Mills to do the following—during a record sales year:

- Improve inventory turns by 20%.
- Increase resource efficiency in the forecasting process.
- Reduce forecast error in one division by close to 50%.
- Achieve a record service level of 99.48%.
- Improve data accuracy and visibility throughout the business [Continental Mills, 2015].

Case 4: Radisys Improves S&OP Process, Supporting Tools Rapidly After Outsourcing Strategy Leads to Real Challenges

Challenge

Radisys, which makes embedded systems and related technology, is located in Hillsboro, Oregon. The company had the beginnings of an S&OP process with limited technology support and was managing to get along until it outsourced manufacturing operations in 2009. This outsourcing resulted in a lack of visibility, and even simple decisions became difficult for the company.

While the contract manufacturer had supply information and Radisys had demand information, they were stored in different places. Radisys also had an outdated, home-grown demand planning software system that faced many limitations and problems. The demand planning tool could only generate a monthly forecast, while the contract manufacturer was operating on a weekly schedule and needed to commit to customers daily. So while the outsourcing strategy was supposed to result in lower costs, it had created more problems and created problems for the whole company.

Approach

Radisys really needed to have supply and demand, including contract manufacturers, in one system. As a solution, the company settled on licensing Steelwedge software for demand planning and S&OP and icon-scm for demand management and order prioritization. Steelwedge is a cloud-based integrated business planning platform, with an easy-to-configure and easy-to-use interface and platform. icon-scm helps Radisys understand the impact of any decision to its own company, to its customers, and to its suppliers. It provides supply chain planning and supply chain collaboration planning solutions that maximize visibility and minimize response time.

Results

Implementing Steelwedge and icon-scm allowed Radisys to manage more by exception, which freed up staff for more value-added tasks. Staff now spend 80% of their time on analytics rather than on basic tasks, as they did previously.

Radisys has realized the following benefits from this new strategy:

- Improvements in forecasting, demand and supply management, and the S&OP process
- Realization of cost advantages from the initial outsourcing strategy
- A single plan that works across demand, supply, and the contract manufacturer

- Integration of the supply chain plan with the company's financial plan, resulting in more of an integrated business planning environment beyond basic S&OP

- A reduction in forecast cycle time from 2.5 weeks to 1.5 weeks

Radisys can now react quickly to market opportunities and demand, which is extremely important in the technology world [*Supply Chain Digest*, 2013].

Continuing to following the SCOR model, in Part III, "Source," we look at a component of the model that can have a huge impact on a company's cost and efficiency; purchasing and logistics costs can range from 50% to 70% of a company's sales dollars.

PART III

Source

9

Material Requirements Planning (MRP)

Procurement and Purchasing Defined

Acquiring materials is the next logical step after the planning process is complete as it is the net result of the planning process, whether for raw materials and components for manufacturing or finished goods for wholesalers, distributors, or retailers.

Because purchased materials, components, and services make up a great deal of the supply chain spending for most organizations, it is a very visible and important component of supply chain management. (Chapter 1, "Lean Supply Chain and Technology: A Perfect Combination," discusses this leverage effect.)

Purchasing is a basic function in most organizations, and for the purposes of this book is defined as the transactional function of buying products and services. In a business setting, this commonly involves the placement and processing of a purchase order. We use this definition for purchasing to avoid confusion with two other frequently used concepts:

- **Procurement (also known as supply management or sourcing)**—The management of a broad range of processes associated with a firm's need to acquire goods and services in a legal and ethical manner to manufacture a product (direct) or to operate the organization (indirect), the foundation of which is provided by the purchasing function.

- **Strategic sourcing**—A process that takes the procurement process further, focusing more on supply chain impacts of procurement and purchasing decisions, and works cross-functionally within the business firm to help achieve the organization's overall business goals. This includes analyzing the company's annual (or more frequent) "spend" with suppliers and supply markets and helping to develop a sourcing strategy that both supports the overall business strategy and minimizes cost and risk.

The purchasing process typically varies from one organization (and industry) to another, but there are some common key elements, as described in the following sections.

The Purchasing Process

The purchasing process starts with a demand or with requirements—for a physical part (inventory) or a service, which will be covered in this chapter. Figure 9.1 shows the purchasing process document flow steps. These steps are generally components in a purchasing (and larger procurement) system.

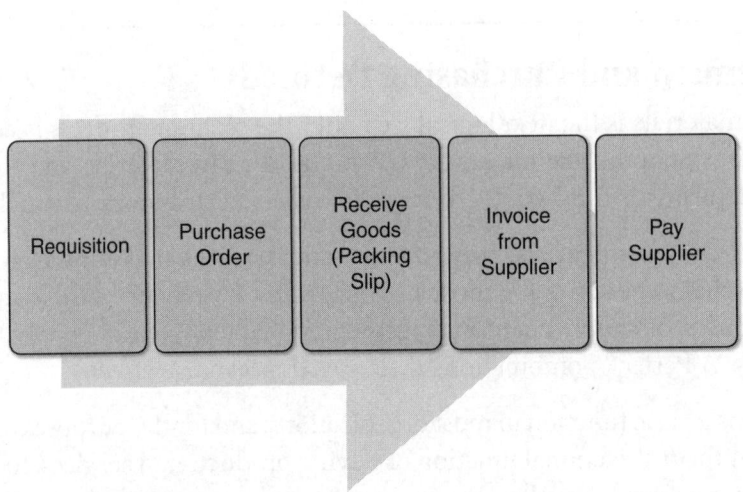

Figure 9.1 Purchasing Process Document Flow

Purchase orders (POs) can be of many types, including the following:

- **Standard**—A standard PO is created when you know the item, price, delivery schedule, and payment terms, typically for a one-time buy.

- **Planned**—A planned PO is created when you are not sure about the required delivery schedules. You are sure about the other details, though, such as the item, price, delivery schedule, and payment terms.

- **Blanket**—A blanket PO is created when you are not sure about the quantity, price, and required delivery schedule. The exact quantity, delivery schedule, and final price will be determined for the supplier by creating blanket releases against the blanket purchase order.

- **Contract**—A contract PO is created when you do not know even the item that is to be purchased. The only information provided on this type of PO is supplier, supplier site, payment terms, and agreement control details. Standard purchase orders are then created by referring to the contract purchase agreement when something is to be purchased against it.

Purchase orders include terms and conditions, which are the contractual agreement part of the transaction. The supplier delivers the products or services, and the customer records the delivery (in many cases after an inspection process). An invoice is sent by the supplier, and it is cross-referenced with the purchase order and any receiving and inspection documents. Payment is then made to the supplier, based on payment terms in the original purchase order.

Types of Business Purchasing

In business organizations, there are generally two types of purchasing activities:

- **Mercantile purchasing**—Arranged by an intermediary for the purpose of resale to meet others requirements. Agents, wholesalers, and retailers are in this category and provide their own channels of distribution to the consumer.

- **Industrial purchasing**—The purchaser is buying to convert material into finished product. This type of purchasing entails buying raw materials, components, supplies and consumables, spare parts and tools, machines, and equipment and office appliance. Industrial purchasing is often split into two categories, direct and indirect, depending on the consumption purposes of the acquired goods and services.

Direct procurement is production-related procurement, and indirect procurement is non-production-related procurement, as described in Chapter 4, "Software and Hardware Sourcing Process and Applications of Supply Chain and Logistics Management Technology".

This chapter focuses on the process and technology for short- to medium-term industrial purchasing requirements. In manufacturing, a master production schedule (or procurement plan, in the case of a distributor or wholesaler) can be used to determine requirements for end or finished goods items (also known as "independent demand"), as discussed in Chapter 7, "Master Production Scheduling." This chapter covers material requirements planning (MRP), the most commonly used tool for calculating dependent demand (i.e., raw materials, parts, components, subassemblies, and so on that make up the finished good or end item).

Material Requirements Planning Described

Once the MPS has been solidified for independent demand items, it can then be "exploded" through a bill of materials (BOM) file to determine raw material and component (i.e., dependent demand) requirements.

The information needed to run an MRP model includes the MPS, a bill of materials (which is similar to a recipe with potentially many levels to it), inventory balances, lead times, and scheduled receipts (i.e., purchase orders, production work orders). All these inputs need to be accurate and up to date. Otherwise, it's the old "garbage in, garbage out" situation, resulting in poor execution and ultimately customer dissatisfaction.

The mechanics of the MPS and MRP systems are basically the same, with the requirements from the MPS (independent demand) driving MRP requirements (dependent demand) via the BOM file.

The information such as forecasts, inventory balances, outstanding production work orders, and purchase orders is netted to create new planned receipts, usually in weekly or possibly monthly time buckets, well into the future, with lead times for dependent demand items offset to create replenishment requirements known as planned orders.

Using a bicycle example, Figure 9.2 illustrates a basic calculation with gross requirements (in MPS, "gross" is the forecast "consumed by open customer orders" in each time period) for the production of 75 bikes of a specific model in week 8. Typically, safety stock or safety time targets would be in place for independent demand items, but for the sake of simplicity, there is none in the example. Because there are 50 bikes in inventory, an additional 25 units of "net" requirements must be produced by week 8. This means 50 wheels and 25 frames must be available in week 6, after offsetting the components' lead time, for the bike production. Through the BOM "explosion," these requirements show up as gross requirements for the wheels and frames in MRP. The same "netting" calculations are then performed to create planned receipts and planned orders for the wheels and frames (and then level 2, level 3, and so on items).

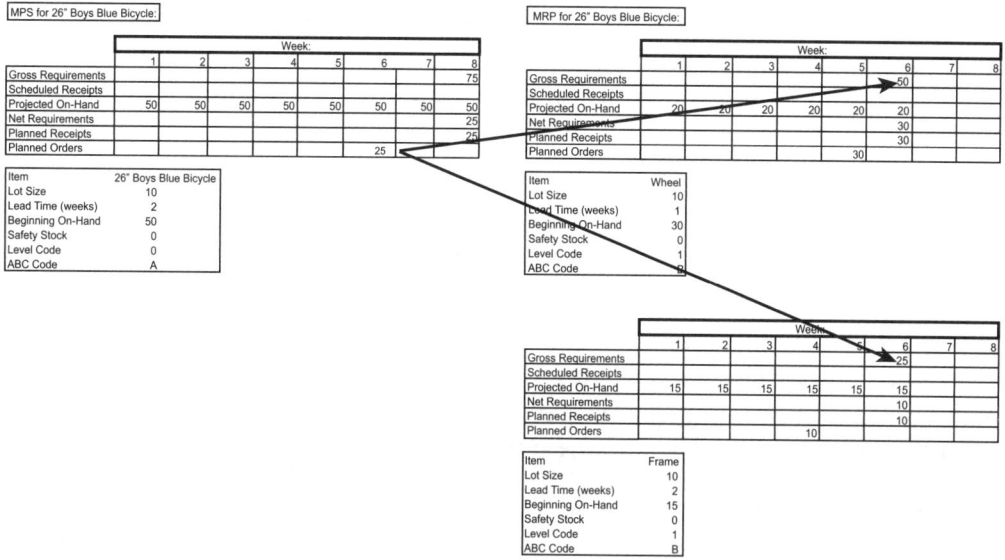

Figure 9.2 MPS and MRP Mechanics

Although it has been said that no safety stock or safety time are required for raw or components, since they are factored into finished goods requirements, the reality is that quality and other issues may arise, along with vendor minimum order quantities, that may call for safety stock.

The actual quantity required is typically rounded up, based on various lot sizing techniques. They range from "lot for lot" (i.e., exact requirements, no matter how small), which is appropriate for just-in-time (JIT) operations to economic order quantity (EOQ) calculations and beyond.

For slow-moving items, an "order time" may be used, which basically states that the planned orders will be grouped together so that one larger order will be placed instead of many frequent small orders. In the case of purchased material or parts, vendors may set order minimums (which can always be negotiated). While this may result in greater holding costs, in the case of slower-moving items, it may be the right thing to do.

Note that in the case of both MRP and DRP (discussed in Chapter 14, "Distribution Requirements Planning [DRP]"), the *R* stands for *requirements*, but there are *resource* versions as well that look beyond material requirements and consider other resources impacted, such as labor, facilities, and equipment. Some, known as closed-loop systems, allow for the planners to schedule work based on period capacity constraints, using smoothing tools that allow the system (manually or automatically) to move requirements around to meet capacity based on priority rules set by the planner, such as order splitting (running parts of a work order at two different times) and overlapping (part of a work order moving to a second operation while the rest is still on the first operation).

Some of these closed-loop systems include capacity requirements planning (CRP) functionality, which generates a more detailed capacity view that is generated in the previously mentioned RCCP system. CRP is performed only after each MRP run, after planned and actual manufacturing orders from the shop floor control system have been considered. This generates a detailed view of what capacity is needed for each work center. The capacity required is compared to the available capacity and identifies over/under load conditions at the work center level.

In reality, CRP isn't really used for interactive planning but more as a verification tool, and it is actually unconstrained or infinite, as it doesn't take into account capacity constraints of each machine or work center.

Note that by using CPP (and, at a higher level, RCCP), much unnecessary waste can be avoided in terms of planning staffing hours and issues created by creating unrealistic plans that then require "firefighting," resulting in expediting, shortages, unplanned overtime, and so on.

In the end, planned orders for both independent and dependent demand are then used (either manually or sent electronically to either an ERP or accounting system) to create production work orders and purchase orders in what is known as *short-term scheduling* (covered in Chapter 13, "Short-Term Scheduling").

Lean and Material Requirements Planning

MRP has been around since the 1970s. It was created by Joseph Orlicky of IBM for planning and time phasing the procurement of parts and materials (dependent demand) tied to the forecasted (independent demand) requirements from a master production schedule. It was the first use of computers to generate the supply orders, which can be quite a task as many finished goods items can be made from hundreds if not thousands of part, components, subassemblies, and raw materials. This was great for large manufacturing companies as it saved a lot of time that had been spent on manual order point systems. MRP was developed at a time when most companies used a demand "push" mentality well before today's customers began expecting shorter lead times and smaller quantities.

Lean-type thinking became prevalent in the 1980s and focused operational management on timing production rates to actual market consumption. It uses replenishment signals known as "kanbans" that summarize customer demand up the supply chain; as discussed previously, this is known as a demand "pull" system. Lean processes can have trouble dealing with extreme variability and can be difficult to use in mixed-mode manufacturing environments with resources shared across multiple products that have unique demand rates. Lean can also lack visibility into the actual flow of materials through the supply chain (especially if it is focused only on the manufacturing operation).

In general, typical thinking on MRP versus JIT is that MRP is usually best for companies that have many product options, frequent engineering changes, and a variable product system, and JIT is best used in environments with fewer product options, fewer engineering and product mix changes, and less variability in demand levels. The reason for this type of thinking is that MRP and Lean view inventory a little differently. MRP typically uses finished goods inventory to meet future customer orders, while Lean attempts to produce on an as-needed basis—just in time—meaning producing to order, not to stock.

Not surprisingly, MRP and Lean also look at the production process differently. Companies using MRP try to steer clear, to some degree, of small lots of a finished product because of cost-inefficiencies (i.e., they don't have Lean processes). Manufacturers that use Lean concepts, however, are more open to small order batches as they have minimal machine setup times.

In the white paper "Lean Finds a Friend in Demand Driven MRP," Carol Ptak and Chad Smith [2011] found that, indeed, MRP and Lean *can* work together, despite the fact that each has its own objectives:

- The objective of MRP is visibility to the total requirements and status picture across the enterprise.
- The objective of Lean is alignment of efforts and resources as close as possible to actual demand.

Both also have weaknesses:

- Lean tends to rely on replenishment kanbans that aren't connected (or visible) to the plant or the enterprise and supply chain level.
- MRP, as it uses complex rules for demand and supply orders, can impede flow.

Ptak and Smith suggest that users should have the ability to define where to place critical "decoupling points" that dampen variability, compress lead times, and minimize working capital. Priority signals can then be used to launch and manage the replenishment of these positions based on sales orders instead of forecasted orders.

When combining the two objectives of visibility and alignment, both of which are conducive to flow, MRP software companies and users can find common ground with Lean manufacturing proponents.

Before reaching that common ground, however, there are a few ways that MRP software needs to change as well.

For example, the article "How Manufacturing Software Can Adjust to Lean Principles" says that "proponents of MRP software believe that today's complex manufacturing challenges require formal planning tools to get an accurate picture of the production requirements. Lean advocates, on the other hand, argue that these planning tools actually get in the way of accurate planning because they're too slow and transaction-intensive to pace to actual consumption, or adjust to demand fluctuations" [Abila, 2012].

The article "described three main ways that manufacturing software can evolve to adapt to the demands of lean manufacturing. Each way focuses on bringing lean principles front and center of manufacturing software packages."

1. **Make Value Stream Mapping a Core Software Component**

   ...It is important to know how information and material flow through a facility in order to help identify and eliminate bottlenecks.

2. **Monitor Cycle Times Intensely**

   Manufacturing cycle times, which are critical measurement in a Lean environment, must be monitored and tracked in order to improve them. This includes the measurement of production status and capacity utilization.

3. **Locate Key Places to Add or Remove Inventory**

   Manufacturing software must not only show what and how much inventory to stock, but also where to stock it, both within the plant as well as throughout the entire supply chain. This helps to better manage volatility and to avoid inventory shortages. [Shmula, 2012]

W.C. Benton and Hojung Shin (1998) also discuss the integration of MRP and JIT (used synonymously with Lean in this and many other cases). They hypothesize that this integration is a result of the evolution of the production planning system used to support JIT implementation in the United States (as well as the implementation of MRP in Japan) to take advantage of both systems and improve overall performance.

Benton and Shin say that three factors have contributed to this "hybrid" type manufacturing environment:

- A multitude of operating problems found when implementing JIT manufacturing techniques
- Organizations (and researchers) better understanding the compatibility between the two systems
- The combination of MRP flexibility in long-term capacity planning and JIT's agility in daily production

In totality, the above-mentioned papers point out the fact that MRP can be used in conjunction with Lean but has to be linked through manufacturing execution systems, as MRP is more for planning. Rather than change MRP to be a Lean system, it must be connected (and configured) to what's actually going on in the shop in terms of Lean tools, such as lot size and lead time reductions, kanbans, and so on.

Material Requirements Planning Technology

While some small to medium-size enterprises (SMEs) may use spreadsheets or affordable standalone MRP systems, such as the one offered by E-Z MRP (www. ezmrp.com; see Figure 9.3), most enterprise resource planning (ERP) systems from companies such as SAP, Oracle, and so on have MRP (and usually RCCP and CRP) functionality.

When ERP systems were developed in the 1990s, they were extensions of MRP systems, tying in customers and suppliers, allowing automation and integration of many business processes, sharing common databases and business practices, producing information in real time, and coordinating business from supplier evaluation to customer invoicing. Today, ERP modules typically include functionality for basic MRP, finance, human resources, supply chain management, customer relationship management (CRM), and even sustainability.

Since the 1990s, ERP systems, through internal development and/or acquisition, have expanded their functionality to include other applications, such as supply chain (forecasting, warehouse management systems, and so on) applications, S&OP, and tighter integration with manufacturing execution systems (MES). Originally, they

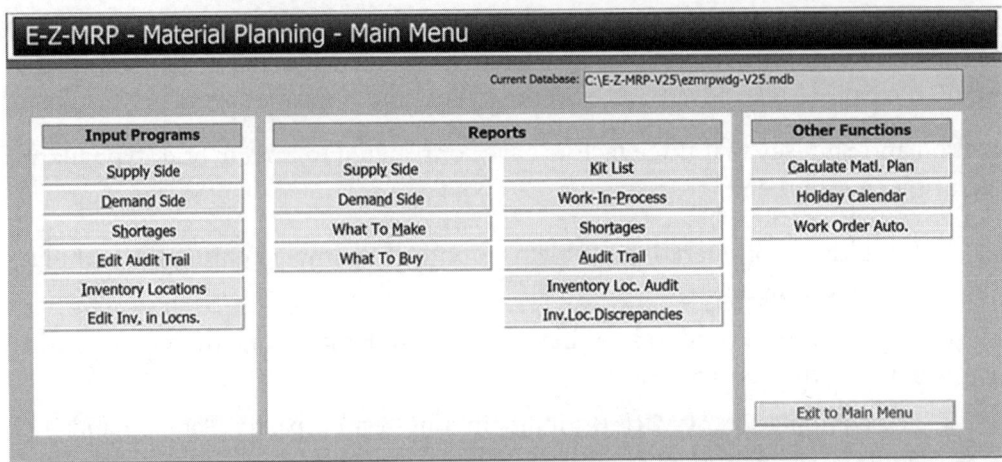

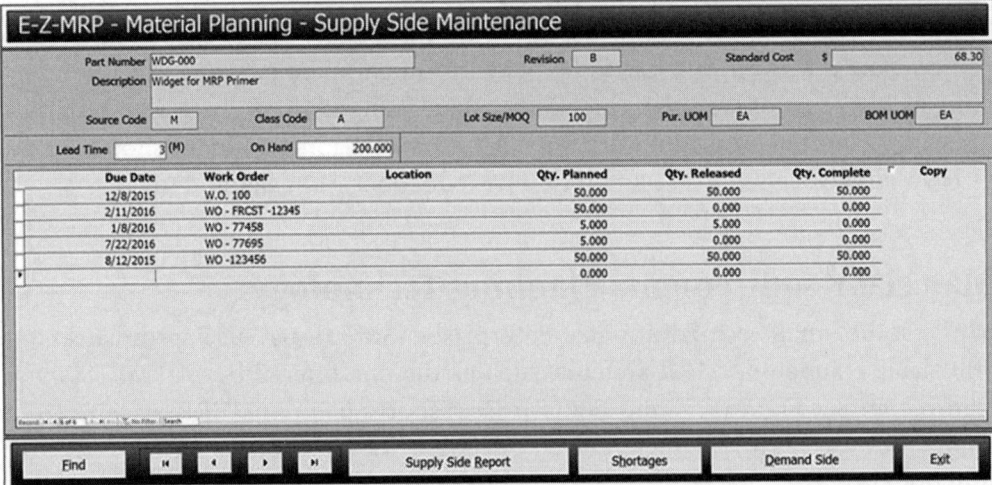

Figure 9.3 MRP Material Planning—Main Menu and Material Planning Screens (Printed with Permission from Weeks Software Solutions, LLC)

were geared toward large manufacturing companies, but they are now available for companies in all industries and of all sizes.

While the basic functionality of MRP technology hasn't changed much over the years, the technology itself has changed and continues to change with the times. Mae Kowalke pointed out five ways that MRP technology is changing:

1. **Better resilience**—With longer supply chains and the need to adapt more quickly at the same time, MRP is increasingly stressing automation and automatic reconfiguration in the face of potential disruptions....

2. **Increased data sharing**—Every engineer knows that the more complex a system, the more points for potential failure. With supply chains stretching in more directions, effectively managing material resources without surprises is requiring firms to better coordinate and share data among each other.

   Beer manufacturer, Heineken, …[brought] supplier data in house…and integrate[ed] it with its MRP systems as if the suppliers were directly part of the business. …

3. **Smarter analysis**—The good news is that there's more data for intelligent MRP, but the bad news is that it can be a challenge to corral all that data and make sense of it.

   Enter the emerging field of big data analytics, which is helping firms pull together the raw data coming from consumers, manufacturing facilities and suppliers.…

4. **Self-reporting inventory**—Tapping into another major trend, MRP is starting to leverage the Internet-of-things (IoT), which delivers automation and self-reporting to materials resource planning for added efficiency and better resiliency. …

5. **Easier implementation**—No trends story would be complete without examining the impact of the cloud. In the case of MRP, the cloud doesn't add functionality as much as it enables fast MRP systems deployment and reduced costs. [Kowalke, 2015]

Lean MRP and Technology Case Studies

The following sections provide some actual examples of companies that have used MRP technology to reduce waste in their production processes.

Case 1: Gables Engineering Moves to a "Real" MRP System

Challenge

Gables Engineering is an avionics manufacturer that designs builds custom cockpit controls, including the switches, housings, and LCD display modules for the airline and airframe industry.

Gables had a pseudo-material requirements planning (MRP) system based on reorder points. The system used current customer demand as well as historical demand to determine what it needed to buy. It was basically moving forward by looking backward (assuming that the past will repeat itself, which we know isn't always the case). The company wanted a product with a Windows front end that could use an Oracle database and could seamlessly interface with other third-party systems.

Approach

Gables formed a team to look at its business processes to see if they needed a new enterprise application system. The team sent a request for quotation (RFQ) to a number of software vendors.

Gables selected IFS enterprise resources planning (ERP)/e-business, which includes a true MRP module to provide a better understanding of actual demand. Previously at Gables, demand was determined by customer orders as well as by a factor for historical demand, causing the company to buy and build unnecessary parts and load shops with unnecessary work and inflated purchased inventories.

Results

Once Gables implemented the new IFS system, it was able to calculate demand using actual customer orders and forecasts. This meant Gables could order and build parts by looking at forecasts rather than at history. This reduced work in process by 50% and overall inventory by 30%. Whereas before the IFS system it had taken two days to get a spare part shipped out, from time of order to time of shipment, now same-day shipping is possible.

The software also tracks the history of changes made to a product down to serial number, including the parts removed from inventory to manufacture it. During a customer audit, Gables was able to show the traceability of a component by using the audited part's subassembly, which it never could have done previously.

Gables has also reduced kitting time to less than one month before assembly, kitting almost 99% of the entire product right before assembling it. It also now has notebooks that allow stock room personnel to review pick lists online and pick and update inventory instantly [Gables Engineering, 2016].

Case 2: Transport Data Systems—A Picture-Perfect MRP Implementation Helps Traffic Enforcement Camera Maker to Profitability

Challenge

Transport Data Systems (TDS), a $2 million company founded in 1995 and located in San Diego, manufactures license plate recognition and capture systems (cameras) for toll roads, parking lot security, and traffic violation processing and enforcement.

TDS never knew how much stock it had of anything, and increasing demand for TDS products drove the company to look for an MRP system. At the time, workers would make a spreadsheet for each job. The spreadsheet included every required part and cost, and for some parts the company simply guessed about costs; in some cases, TDS assumed it had inexpensive parts on hand until build time, when TDS found that the parts weren't there.

Results

Once TDS had licensed E-Z-MRP, the vice president of operations was freed up from administrative tasks and could focus on product improvements, strategic projects, and other aspects of his work that only he could do.

The new software also allowed TDS to determine a selling price quickly instead of recalculating from each new spreadsheet how much it would cost to build the product. The MRP system is now consistently being used as a standard operating platform, no matter who used the MRP system helping achieve stability and accuracy in the process [www.e-z-mrp.com, 2016].

Case 3: Raytheon Streamlines and Automates Its Material Requirement Planning Processes with Exostar's Supply Chain Platform

Challenge

Raytheon is a technology and innovation leader that specializes in defense, homeland security, and other government markets throughout the world and provides electronics, mission systems integration, and other capabilities, as well as a broad range of mission support services.

Raytheon divisions relied on a variety of MRP processes and "homegrown" and packaged software systems for the delivery of products and services to customers. With those systems, it faced the following issues:

- **Siloed processes and systems**—Each of Raytheon's six business units had implemented its own MRP solutions.
- **Manually intensive interactions**—The MRP systems couldn't be accessed directly by suppliers, so all scheduling communications required an excess of manual intervention, resulting in increased cost, time, and risk.

Approach

Raytheon decided to transition to a collaborative MRP (cMRP) supply chain platform (SCP) solution from Exostar that would deliver the following benefits:

- Automate processes as much as possible.
- Leverage existing infrastructure to minimize impact and cost.
- Enable increased collaboration between buyers and suppliers.
- Provide standardization and integration across all existing MRP systems.

Results

Raytheon's Integrated Defense Systems (IDS) group was the first business unit to transition to SCP and connect its existing MRP systems with SCP. The IDS group has been able to modify its existing MRP process to create a leaner, more automated, standardized, and collaborative process that reduces manual processes, increases productivity, and increases integration with suppliers.

Overall, Raytheon anticipates that after full implementation at all business units, it will see savings of up to $3 million per year, thanks to the following improvements:

- **Increased reliability**—On-time delivery of goods from suppliers is expected to increase by 10% or more.
- **Streamlined process**—New system should reduce manual reentry of information, printing/faxing, and phone/email communications.
- **Optimized resources**—The software should make it possible for Raytheon to move 12 to 15 employees per business unit from administrative to more high-value tasks.
- **Reduced supply chain risk**—Raytheon expects better performance visibility and exception management.
- **Improved consistency**—Raytheon plans to implement the cMRP solution across all six Raytheon business units over time [Raytheon, 2013].

In Chapter 10, we will examine the use of technology in a Lean procurement process, which tends to be a bit more strategic and tactical in nature than the subset of purchasing—which is often tied directly to MRP systems as described in this chapter.

10

Procurement (and e-Procurement) Systems

The Procurement Process

Today, technology is heavily used in the short- to medium-term purchasing process in a variety of sources, such as an MPS, procurement planning, and MRP. It is also used for the broader, longer-term needs of an entire procurement/sourcing process (see Figure 10.1).

As described in Chapter 9, "Material Requirements Planning (MRP)," *procurement* (also known as *sourcing and supply management*), the focus of this chapter, is the process of managing a broad range of processes associated with a firm's need to acquire goods and services in a legal and ethical manner in order to manufacture a product (direct items) or to operate the organization (indirect items), the foundation of which is provided by the purchasing function. As shown in Figure 10.1, the procurement process typically includes the functions of determining purchasing specifications, selecting the supplier, negotiating terms and conditions, and issuing and administering purchase orders.

Preparing and managing purchasing documents involved in this process has always been a time-consuming process. Most firms have streamlined the document flow process using Lean and other process improvement techniques to reduce the paperwork and handling required for each purchase.

Automation of Procurement Documents and Processes

As discussed in Chapter 4, "Software and Hardware Sourcing Process and Applications of Supply Chain and Logistics Management Technology," a variety of purchase orders

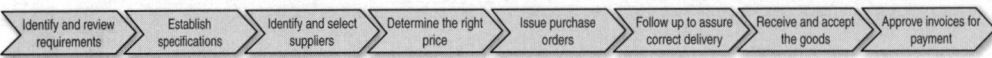

Figure 10.1 The Procurement Process

may be generated as a result of the procurement process (e.g., discrete order, blanket order, etc.).

Numerous other documents and/or information requirements can be automated and integrated in an e-procurement process, including the following:

- **Identifying and reviewing requirements and establishing specifications**— Specification sheet, statement of work, product requirement, customer order/ MRP requirement, and purchase requisition
- **Selecting suppliers (approved)**—Purchasing card, e-catalog, EDI, stock check, and reorder point
- **Selecting suppliers (unapproved)**—Request for quote/information (RFQ/ RFI) and request for proposal (RFP)
- **Issuing purchase orders (PO)**—Purchase order approval and release/ acknowledgement and blanket order
- **Handling delivery and receipt**—Bill of lading (B/L), packing slip, discrepancy report, kanban, and receipt acknowledgement
- **Handling payments**—Supplier invoice, matching PO and invoice, paying invoice, and updating supplier scoreboard

It should come as no surprise that procurement is an area where technology has been heavily applied over the past 25 years. In fact, every one of the steps involved in the procurement process typically utilizes technology to some degree or other.

In its totality, the suite of tools used to achieve efficiency in purchasing transactions is broadly referred to as *e-procurement*. Companies are using e-procurement tools to manage the flow of documents by automating the document generation process and electronically transmitting purchase documents to suppliers.

Lean and Procurement

Supply chain costs can range from 50% to 80% of a company's sales, depending on the industry, so it's not difficult to see why it's an important area to look for waste.

Lean sourcing, or *procurement*, is a different way of looking at and working with suppliers. It involves a greater use of partnerships and alliances as well as a greater need for coordination and collaboration, often utilizing some of the technological tools mentioned previously for accuracy, timeliness, and efficiency.

Traditional supply chains are managed more on a cost basis, negotiating with many suppliers. While this may still be effective in some instances (e.g., commodities), Lean procurement is all about long-term partnering with fewer, longer-term suppliers with less reliance on low-cost bidding. Motorola, for example, has eliminated traditional supplier bidding by adding emphasis on quality and reliability and in some cases may sign contracts that are in place throughout a product's life cycle [Heizer and Render, 2011]. In this way, the relationship can be mutually beneficial. The value is created by economies of scale and long-term improvements (see Table 10.1).

Table 10.1   Lean Supply Chain Characteristics

Characteristic	Traditional Supply Chain	Lean Supply Chain
Suppliers	Many	Few
Interactions	Confrontational	Collaborative
Relationship focus	Transactional	Long term
Primary selection criteria	Price	Performance
Length of contract	Short term	Long term
Future pricing	Increased	Decreased
Lead times	Long	Short
Order quantities	Large lots	Small lots
Quality	Expensive inspection	Quality at the source
Inventory (supplier and customer)	Large	Minimal
Information flow	One way	Two way
Flexibility	Low	High
Product development role	Small	Large (collaborative)
Trust	Limited	Extensive

As a result of this type of relationship, where trust is very important, suppliers are more willing to get involved in just-in-time (JIT) partnerships, share in the design process, and contribute technological expertise. For example, when Cessna Aircraft opened a new plant in Kansas, it set up consignment and vendor-managed inventory programs with some select suppliers. One supplier, Honeywell, was allowed to maintain avionic parts onsite. Other vendors that participated kept parts at a nearby warehouse to supply the production line on a daily basis. This was a win–win situation as Cessna was able to execute JIT inventory replenishment for parts, and suppliers gained better insight into Cessna's production requirements and could offer suggestions for product improvements, thereby strengthening the relationship [Heizer and Render, 2011].

Some suppliers may be somewhat hesitant because of issues such as relying too heavily on one customer, shorter lead times, and smaller order quantities. In a true partnership, the customer must be willing to work with the supplier and share costs, training, and expertise so that they're not just "passing off their problems upstream." Of course, a company needs to always have a backup plan and single source (i.e., one supplier for an item) only where there is very little risk involved (e.g., commodity-type item, easily substituted part).

There are many Lean opportunities in procurement, including the following:

- **JIT, as in the Cessna example above**—There may also be a potential application for vendor-managed inventory (VMI), where a supplier manages its customer's inventory of parts and supplies (see Chapter 19, "Collaborative Supply Chain Systems").
- **Batch size and lead time reduction**—Producing smaller quantities of items more frequently reduces inventory and cycle time.
- **Blanket orders**—A customer may place a single purchase order with a supplier, containing multiple delivery dates scheduled over a period of time, in many cases at predetermined prices.

It is often said that "if you can't measure it, you can't improve it." This adage applies to all of the applications mentioned in this book. By performing Lean assessments and supplier reviews, you can determine how lean your supplier is and what progress has been made toward that goal.

Besides process improvement itself in the procurement process, the benefits of electronically generating and transmitting purchasing-related documents include the following:

- A virtual elimination of paperwork and paperwork handling
- A reduction in the time between need recognition and the release and receipt of an order
- Improved communication both within the company and with suppliers
- A reduction in errors

- Lower overhead costs in the purchasing area
- Less purchasing personnel time spent on processing of purchase orders and invoices and more time on strategic value-added purchasing activities

Procurement Technology

A variety of technologies are available today to help an organization automate and improve its various procurement processes. For the purposes of this book, we can define e-procurement as business-to-business purchase and sale of supplies and services over the Internet, which can be integrated with internal computerized procurement processes and systems, as identified in Figure 10.1.

In the area of procurement, which may sometimes include e-procurement functionality, there are two types of software vendors:

- Enterprise resource planning (ERP) providers offering both internal procurement (including individual MRP vendors; see the example in Figure 10.2) and e-procurement as one or part of their modules
- Services or vendors focused specifically on e-procurement

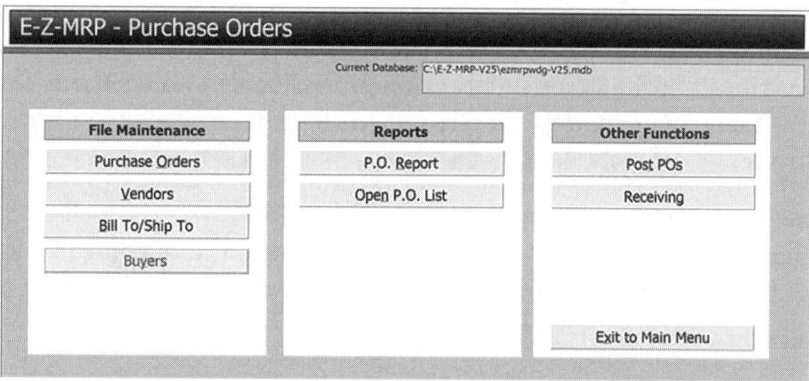

Figure 10.2 MRP Material Planning—Purchase Order Main Screen (Printed with Permission from Weeks Software Solutions, LLC)

Procurement software is a computer program or suite of products that helps to automate (and thereby improve) the processes of purchasing materials and inventory maintenance of goods. Following the typical procurement process, it can generate purchase orders, execute the ordering process online, match invoices to materials received, and pay bills electronically. Again, more often than not, systems today include e-procurement functionality and also leverage the benefits of the Internet. As a result, the benefits of using procurement software include ease of administration and potential cost savings as well as having a single interface for procurement to monitor company spending.

Procurement software helps to efficiently manage a variety of activities, including the following:

- Creating a purchase order based on need
- Verifying a purchase order
- Submitting a pending purchase order for approval or rejection
- Automating an electronic purchase order transmission
- Confirming or cancelling purchase orders
- Executing financial and inventory transactions when ordered materials arrive
- Gathering and analyzing data and to improve profitability
- Streamlining and standardizing administration (For example, procurement systems generally offer multi-currency support as well as tools that can automate purchases and purchasing approvals.)

These systems can also connect users with large networks of qualified suppliers, which is a critical capability for supply chain professionals who are trying to identify the most reliable raw materials suppliers at the best price, wherever they might be sourced from.

As opposed to procurement modules internal to ERP systems such as SAP and Oracle, there are also "standalone" procurement solutions that come in a variety of forms. Charles Dominick identified 10 types of standalone procurement software systems:

- **Spend analysis**—Allows you to find purchasing patterns within categories, such as by suppliers, that might offer cost savings, performance improvements, and overall efficiencies. Most spend analysis vendors have been acquired by other types of procurement software vendors.
- **Supplier discovery**—Allows you to search for suppliers that meet specific criteria, such as capabilities, location, and supplier diversity. Examples include ThomasNet and Ariba.

- **Supplier information management**—Allows you to efficiently collect and maintain accurate supplier information, including contact information and certification status, directly from suppliers. Examples include HICX Solutions and Hiperos.

- **eSourcing**—Allows you to get quotes and proposals electronically from suppliers quickly. This can be done privately or can allow suppliers see their rank among bidders to increase competitive pressure. Examples include WhyAbe, K2Sourcing, and Trade Extensions.

- **Contract management**—Enables you to prepare contracts using various templates. It can electronically route contracts for approval, track revisions, notify of contract expirations, and store executed contracts. Examples include Selectica and Prodagio.

- **EProcurement**—Allows end users to search catalogs of preapproved products and services from contracted suppliers, create requisitions, and have some requisitions turned into purchase orders (either manually by buyers or automatically). Examples include ePlus and eBid.

- **E-invoicing (also known as ePayment)**—Enables you to efficiently receive accurate supplier invoices electronically. May include "dynamic discounting," which allows suppliers to reduce the amount your organization owes in exchange for faster payment. Examples include Taulia and Tradeshift.

- **Supplier management**—Enables you to track and/or rate supplier performance using manual or scorecard-style ratings. Also integrates with other systems to gather actual performance. May include risk assessment capabilities. Examples include Aravo and BravoSolution.

- **Combination solutions**—Allows for direct payment to suppliers, commonly referred to as *procure-to-pay* (*P2P*). Those that offer the option of soliciting quotes from suppliers are called *source-to-pay solutions*. Examples include Coupa and Puridiom.

- **Complete suites**—Some software vendors have suites of many or even all of the solutions listed above. Examples include GEP, Zycus, SciQuest, and iValua [Dominick, 2015].

Lean Procurement and Technology Case Studies

The following sections provide some actual examples of companies that have used technology to improve their procurement processes.

Case 1: Enabling Online Supplier Collaboration at Toshiba Semiconductor Company

Challenge

Toshiba Semiconductor Company wanted real-time information for its global operations in order to stay a global leader in the industry. Lacking that information would limit the company's success in the future. At the time, purchasing employees bought products locally, and there was no sharing of information among buyers, factories, or headquarters; these activities were being conducted separately primarily because no central database existed.

Approach

Toshiba selected JDA software for spend optimization to help manage its supplier relationships using a web interface with its customers and suppliers. This would enable Toshiba to operate on a real-time basis, using accurate information, so Toshiba and its suppliers could collaborate on sourcing and procurement for supply management. In this way, the company could integrate product development, sourcing, supply planning, and procurement across the entire supply chain.

Toshiba decided to implement JDA Negotiate and Strategic Sourcing for direct materials and information-gathering and decision-making processes. This would enable the company to send out RFQs to suppliers via the Internet and help create a supplier database shared by all the purchasing staff to assist in the selection of the best suppliers in future negotiations and to make balanced scorecards for each supplier.

Results

As a result of its successful implementation, JDA solutions enabled Toshiba to gain competitive advantage by refining its supplier base and adding speed, efficiency, and reliability to purchasing. The company can now handle between 7,000 and 8,000 RFQs per site at six of its major factories in Japan.

Toshiba feels that it has achieved a competitive edge due to the increased speed and intelligent decision making available through the JDA software. It expects to be able to reduce the number of preferred suppliers by using a balanced scorecard with information contained in the new database. It also expects purchasing agents to become more strategic because they can collaborate with product designers in the design stage, where 80% of a product's cost is determined. [JDA, 2016]

Case 2: Clariant—Increasing Interenterprise Productivity and Extending Its SAP Software Investment Value

Challenge

Clariant, a global leader in specialty chemicals, markets innovative chemicals in a variety of business areas. The company wanted to improve the accuracy of its catalogs for its global supply base and to develop more collaborative supplier relationships. In addition, it wanted to improve invoice cycle times.

Approach

By deploying Ariba Procurement Content, PO Automation, and Invoice Automation solutions, Clariant determined that it would be able to purchase all indirect goods and services through its existing SAP ERP software system.

Results

Clariant eventually deployed the Ariba Procurement Content solution to manage more than 300 catalogs, deployed the Ariba PO Automation solution in 21 countries, and rolled out the Ariba Invoice Automation solution in Germany and Switzerland. This integrated smoothly with the existing SAP Supplier Relationship Management application for order initiation. It enabled a consumer-like shopping experience, covering all countries with one user-friendly solution. To accomplish this, Clariant used Ariba services, which incorporated catalogs and suppliers on the Ariba Network.

Clariant reached its goal of purchasing all indirect goods and services through the integration of Ariba PO and Invoice Automation with its SAP ERP application, thereby increasing order accuracy, reducing non-catalog orders, and streamlining invoice processing in Germany and Switzerland. It allowed procurement personnel to focus on higher-value activities and improved collaboration internally and with suppliers. [SAP, 2016]

Case 3: New Purchase-to-Pay System Allows Smarter Processes at Atea

Challenge

Atea is a leading supplier of IT infrastructure in Europe; it helps enable the smooth running of its customers' IT purchasing, delivery, and service processes by delivering the necessary hardware and solutions.

Aside from its hardware purchases, which are the majority of items purchased and are handled by central purchasing, Atea lets employees do their own purchasing for indirect items. Currently department heads have to provide authorization twice: once to authorize a purchase and later to authorize the invoice after delivery is made.

Approach

Atea looked for a combined technology solution that included purchasing, automated invoice processing, and travel and expense management. It had to support Atea's decentralized (indirect item) purchasing strategy and integrate with the existing ERP and payroll systems, all the while being user friendly.

Atea needed the solution help to optimize its purchasing, invoice, and expense-handling processes as well as integrate the invoice processing and travel and expense management systems. It needed a system that would enable department heads to deal with purchases just once.

Results

Atea licensed a purchasing system from Basware. Employees can now create a purchase requisition and get it approved electronically by their department head. When the invoice arrives, it's already been approved and can be sent for payment automatically, thereby reducing the work of two people by 50%. The following are some of the benefits Atea has realized with this system:

- Atea no longer has to search for invoice documents, ownership of invoices, or approvals, and it is assured that the right purchases are made.

- The purchasing system now integrates with invoice and travel and expense processing and with the entire payables side, enabling many approvals to be granted automatically.

- The responsibility for invoice posting has been delegated to individual departments, with decentralized invoice posting.

- There is no need to send invoices from one department to another anymore, and departments have a better idea of what they are actually spending.

- Atea also chose to shut down expensive manual advances and is switching as many staff as possible to personal liability credit cards [Basware, 2016].

The type of procurement and purchasing systems described in this chapter are typically part of or integrated with enterprise resource planning (ERP) systems, the topic of Chapter 11, "Enterprise Resource Planning (ERP) Systems."

PART IV

Make

11

Enterprise Resource Planning (ERP) Systems

Enterprise resource planning (ERP) systems have grown to incorporate a huge diversity of functionality, including topics covered in other parts of this book. In this chapter we discuss ERP systems from a perspective of enabling production and related internal supply chain and logistics operations processes.

While some may not consider ERP systems supply chain management tools, a great deal of the functionality is supply chain and logistics related (see Figure 11.1), especially when you consider potential add-on modules for forecasting, warehouse management systems, etc. ERP systems were originally an extension of MRP systems, used to integrate all internal processes as well as customers and suppliers. They allow for the automation and integration of many business processes, including finance, accounting, human resources, sales and order entry, raw materials, inventory, purchasing, production scheduling, and shipping, resource, and production planning and customer relationship management. An ERP system shares common databases and business practices to produce information in real time and coordinate business processes ranging from supplier evaluation to customer invoicing.

E-businesses must also keep track of and process a tremendous amount of information; they have realized that much of the information they need, such as stock levels at various warehouses, cost of parts, and projected shipping dates, can be found in their ERP system databases. As a result, a significant part of the online efforts of many e-businesses is adding web access to existing ERP systems.

ERP systems have the potential to reduce transaction costs and increase the speed and accuracy of information but can also be expensive and time-consuming to install.

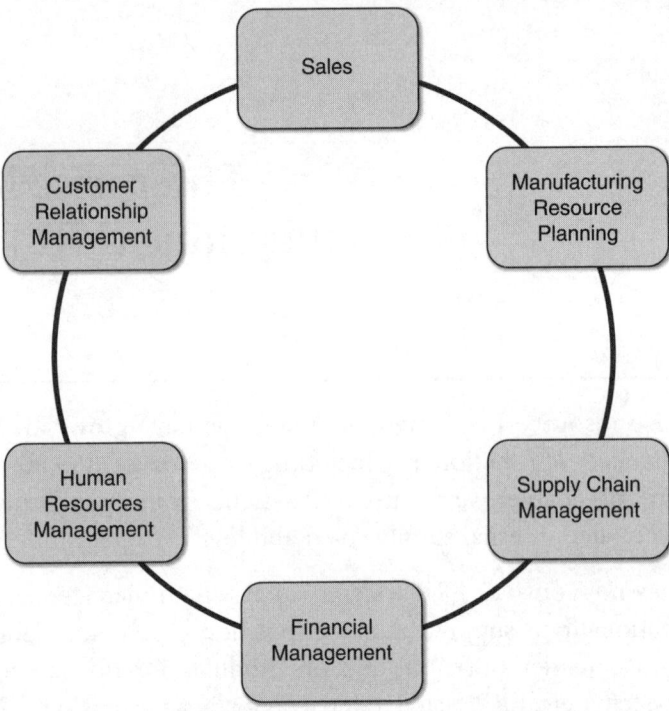

Figure 11.1 Typical Modules in a Basic ERP System

Lean Supply Chain and ERP Systems

Many have called Lean a "pen-and-pencil" type of tool, and some supporters talk about the elimination of ERP systems. However, the reality is that a manufacturer's ERP system, while a necessity, should be simplified as it can include a vast array of day-to-day functionality, including general ledger, accounts payable, purchasing, receiving, and order management, to name a few areas. An ERP system is not only needed to manage existing business practices but also for tracking and analyzing current manufacturing processes to help find areas that need improvement.

In addition, many of today's ERP systems have extended their applications to support basic Lean principles of value definition and specification, value stream mapping, flow, and demand pull. Furthermore, ERP and Lean production offer many of the same benefits to an organization, such as inventory and lead time reduction, quality improvement, and customer service improvement.

As manufacturing industries (and service industries, to a degree) have moved from mass production to Lean production, with the supply chain now heading in that direction as well, it can be said that we are now evolving into a phase of technology-enabled Lean. However, it must be understood that ERP systems support and enable Lean but don't drive it. It's the Lean culture, training, and tools that make it happen. ERP systems support and help enable Lean initiatives.

Many things can be reconfigured or modified in an ERP system to help enable Lean manufacturing and the supply chain. However, before making such changes, it is critical for an organization to resolve any conflicts between manufacturing efficiency and the customer service and sales departments. Typical Lean processes require a shift to smaller batch sizes, lead times, setup times, and so on, which can be in conflict with the revenue goals of sales department, which often like to have excess inventory "just in case." Then these leaner production schedules should be tied to sales projections and actual customer demand, which can be configured in an ERP system. Then metrics and measures can be set in the ERP system for tracking and controlling purposes.

In any manufacturing operation, it is a good idea for an enterprise application to support multiple production strategies, including make-to-order (MTO), make-to-stock (MTS), engineer-to-order (ETO), and configure-to-order (CTO). Even MTS manufacturers have to consider the different levels of demand for various products and part numbers; they should consider having a parallel MRP system that allows stable products to be run MTS and the many volatile, small-volume "C" items to be run in an MTO environment (and/or MTS, with excess safety stock to compensate for poor or nonexistent forecasts). This increases the organization's efficiency on products that it make sense to manufacture in large quantities and its responsiveness on products that are best handled as special orders.

An ERP system should have at least some tools that allow production to become pull based, using processes such as a kanbans—Japanese for "visual signal"—which, based on downstream demand, tell you what to produce and when to produce it. Having this functionality allows companies to pull products through a leaner supply chain while minimizing work in process, maximizing flexibility and responsiveness, and avoiding excess quantities of finished goods.

Enterprise applications should allow for operations at multiple sites because with any complex supply chain, you really are operating in a multisite environment. The more integrated the operations between these different sites are, the more responsive you can be as the information flows seamlessly through your internal supply chain processes as well as customers and suppliers, whether these sites are under your direct control or not. This will allow your company to see multisite functionality as a tool for collaboration throughout the supply chain.

Master data management is critical for companies that have recently been through a merger or acquisition, especially where different parts of the company have multiple part numbers for the same item. Visibility and integration through the supply chain are critical to Lean, but it is hard to have enterprise-wide visibility and integration when you have duplicate data and records for identical parts. Even if you are within a single global company, it is important to have multicurrency and multiple-language support because with any extension of the application to a customer or vendor overseas, you may end up with duplicate data, which hampers your Lean supply chain improvements. In general, you need to have a common language for customer numbers, item codes, and so on to integrate the business and share plans and to bring Lean efficiencies to your supply chain [IFS Software, 2009].

ERP Technology

Enterprise system software is a multibillion-dollar industry that helps support a variety of business functions; it has been the largest category of capital expenditure in U.S. businesses over the past decade or so. While early ERP systems focused on large enterprises, smaller enterprises increasingly use ERP systems to run their businesses in industries such as manufacturing, wholesalers/distributors, health care, government, retail stores, hotels, and financial services.

There are literally hundreds of ERP software vendors, ranging from very large ones with expensive offerings up to millions of dollars (e.g., SAP and Oracle) to midsized/priced and relatively small vendors costing as little as $100,000 and up (e.g., Netsuite and Exact). They can vary in terms of functionality and platforms (e.g., client/server or cloud-based/on-demand/Software-as-a-Service [SAAS]), and they can serve general or only specific industries.

It is beyond the scope of this chapter to get too much further into the details of this technology beyond what we've already covered in terms of basic functionality and impact on the Lean supply chain. Keep in mind, however, that the selection and implementation of these very critical systems can mean success or failure for an organization as speed and accuracy are not only important from a Lean perspective but can offer a company a distinct competitive advantage in today's global economy. In addition, when poorly managed, the selection and implementation can be very costly and wasteful to a company; not only is the software license itself expensive but the total cost of ownership—including training, consulting, technical and maintenance, hardware upgrades, and customization costs—can be three to five times the cost of the software license.

An article in *CIO* magazine offered nine tips for selecting and implementing an ERP system [Schiff, 2014]:

- **Get upper management support**—Lack of upper management support and involvement can lead to resources at lower levels not being dedicated and engaged in the implementation project.

- **Make a clear and extensive list of requirements before you start looking at vendors**—Any good project must start with a definition of its scope. This includes identifying specific business processes and their functional and system requirements. It is critical that you work with end users, IT, and senior management from the start to find an industry-specific ERP system that has the tools and features you need to solve your business requirements. This up-front effort will pay off in the long run.

- **Don't forget mobile users**—Accessing ERP systems from desktops only is no longer an option, so look for an ERP solution that allows users to also connect securely via smart phones and tablets.

- **Carefully evaluate your options before selecting your ERP system**—Make sure you have clear requirements and priorities as well as participation and input from key stakeholders during the evaluation stage to ensure better acceptance and user adoption.

  Reporting and metrics in the selected system are also important. Make sure the existing reports in the system have the metrics you need to drive your business, hiring, and resourcing.

  Integration is important as the system must work with your existing legacy and/or critical office systems. Also, if possible, try to find a vendor that specializes in your industry or at the very least has clients in your industry that you can talk to.

- **Get references**—If possible, ask other customers what went right, what went wrong, and what they might have done differently. You can also network with industry associations you belong to and ask colleagues for ERP recommendations.

- **Think before you customize**—Think about the amount of customization required for the ERP system; keep in mind that the more customization that is required, the higher the cost, not only initially but when you upgrade to new releases. Also understand your tolerance for longer implementation cycles; while turnkey solution may have less flexibility, they will also likely have more stability and less initial and ongoing cost.

Many companies' basic business processes are very similar, such as paying invoices, collecting revenue, and procuring supplies (even though they may not think so!). So there may be an opportunity to take advantage of standard "best practice" processes that have been tested by many other companies.

If a business function believes it has a case for a customization, require justification since the cost of the customization is not only writing and testing the code but providing long-term support of the custom code that may require special handling when you upgrade your software.

- **Factor in change management**—Most ERP projects entail huge change in organizations and impact the culture of the company. So you need to develop control and communication plans and workshops to help with implementation and adoption of the systems.

- **Appoint an internal ERP product champion**—Don't just use a vendor-appointed project manager. It's important to put your best people on the job as a lot is at risk.

- **Provide the necessary time and resources for training on the ERP system**—You should identify department-specific needs up front and allow sufficient time to develop and deliver training programs. Where possible, give current employees the opportunity for more in-depth instruction so they can become expert resources for their fellow employees (i.e., "train the trainer"). This can help reduce the "us versus them" dynamic that often occurs.

Lean Supply Chain and ERP Systems Case Studies

The following sections provide some actual examples of companies that have used ERP systems to improve the accuracy, efficiency and timeliness of the transactional aspects of their businesses.

Case 1: Radio Flyer Teamed with Ultra Consultants on ERP Selection, Business Process Improvement, and Implementation Management

Challenge

Radio Flyer is one of North America's most recognizable wagon and toy manufacturers, best known for its red wagon.

Radio Flyer wanted to standardize its global operations onto a single ERP platform. It was limited by its old AS-400-based ERP system. The company wanted to improve visibility into the supply chain, optimize inventory, increase accuracy of sales forecasting, automate sales and financial operations, and improve reporting with real-time data analytics.

Approach

As is commonly the case, Radio Flyer didn't have in-house ERP expertise, so it brought in Ultra Consultants to help with the system selection and implementation process. Ultra helped Radio Flyer identify gaps between the current state and the desired future state of operations, as well as systems that would help close those gaps. The maps helped show the company the pain points, bottlenecks, and waste in its processes and how to eliminate them.

During the software evaluation process, Radio Flyer focused on key functionality to understand which ERP features and functions would be needed for business process improvement. It also helped to speed up the ERP selection process. Radio Flyer ultimately chose Oracle's JD Edwards EnterpriseOne system.

Results

With the help of an outside consultant, Radio Flyer was able to leverage insight, knowledge, and methodologies to keep its internal team on course. This process helped the company reduce effort and efficiently reach its goals of identifying process gaps; determining which ERP modules, features, and functions would improve its processes; and ultimately choosing and implementing the best solution for the business in the shortest amount of time possible [Ultra Consultants, 2016].

Case 2: Flexpipe Systems Inc. Works with IFS Applications for More Flexibility

Challenge

Flexpipe Systems Inc. manufactures and sells spoolable composite pipeline systems used for oil and gas gathering systems, water disposal, and other applications where a corrosion-resistant, high-pressure pipeline is required. Flexpipe Systems used a heavily customized (and non-upgradable) enterprise resource planning (ERP) system that was handcuffing the company because a lot of the standard functions were beginning to be hampered by changes to the source code.

Approach

In 2008, Flexpipe Systems was anticipating significant future growth requirements, and it wanted a solution with additional functionality, such as the use of recipes that could provide extra flexibility. It determined that IFS Applications could provide the right solution. The IFS ERP system would offer major and measurable efficiencies to the company's inventory management system, enabling Flexpipe to reduce safety stocks and more quickly ship orders.

Results

Before switching to IFS, it took Flexpipe approximately a day and a half to receive inventory from a shop order. Now with IFS, transactions are live within 20 minutes. The company can finish a product, get it on a truck, and ship it out of its facility within 20 minutes of its completion. In the past, this process had typically taken a few days.

Flexpipe had also experienced problems with its weekend production schedule. On Monday mornings it would spend considerable time closing shop orders that had been created during the weekend. With the new ERP system, this process happens instantly and on the shop floor; the system has eliminated the delays that had formerly been caused by receiving three days of production into inventory on Monday morning.

Before adopting IFS, Flexpipe needed to maintain four days of safety stock of its semi-finished inventory parts because it had a weekend of shop orders that had not been closed out. After the implementation of IFS, the company was able to significantly speed up the shop order closing process, and it saw a 60% reduction in safety stock.

Lot tracking for Flexpipe's pipe fitting products is critical. With its old ERP system and initially with IFS, the company was performing extensive non-value-added work—recording inventory and shop floor transactions manually and then entering them into IFS Applications. Shipping or receipt details, including the part number and serial tracking numbers, would then be entered into IFS when someone had time.

After implementation of the new system, Flexpipe was able to leverage IFS Applications's service oriented architecture, using web services to drive data directly through the system's business logic. Handheld devices now read bar codes to capture serial, lot, and batch data and automatically enter it in IFS Applications. Bar code integration has delivered Lean improvements and allowed Flexpipe Systems to reduce the risk recalls [IFS ERP Systems, 2015].

Case 3: Automotive Supplier Nissen Chemitec America Accelerates Lean Operations with IQMS ERP

Challenge

Nissen Chemitec America (NCA), a leading automotive supplier operating in an industry where a large number of suppliers compete for relatively few customers, understands the importance of Lean manufacturing. It had an ERP system that conformed to automotive customers' very strict requirements. However, the system hindered NCA's ability to advance Lean manufacturing principles. The system was designed for suppliers of the Big Three automakers, so it didn't meet many of NCA's needs. It was cumbersome, required heavy data entry, and blocked NCA's efforts to be Lean as it required a lot of maintenance.

NCA began looking for a more tailored ERP solution, one that was built specifically for contract manufacturers serving the automotive industry. The company was looking for a fully automated system that met automotive compliance requirements such as electronic data interchange (EDI), labeling, and quality functions but that was also robust and scalable. It also had to comply with customers' quality standards and business transaction requirements.

Approach

Ultimately NCA selected EnterpriseIQ from IQMS, which has specific functionality designed for the auto supply industry, featuring a combination of manufacturing-specific functionality and automotive industry compliance standards.

NCA's previous ERP system slowed Lean progress in a number of ways. For example, it was built on multiple databases that required repetitive data entry across various modules and functions; it was not automated; and it had limited bar code scanning capabilities. NCA had to do production scheduling manually, using spreadsheets, and production data had to be keyed in separately for reporting. One of the most significant areas of concern for NCA was that while the system supported EDI, the transactions were sent via modem, which delayed the company's ability to find and correct problems before and after they occurred. This slower-than-desired method of transmitting data and supporting customer requirements also cost the company between $3,200 and $3,600 monthly. On top of that, the company was experiencing between 30 and 40 EDI-related shipping errors per month.

Results

After implementing an IQMS ERP system, NCA reduced its monthly EDI cost by 90% (to only $300) and eliminated most shipping errors. In fact, from the moment NCA turned on the IQMS system, it started shipping without errors.

Incoming EDI files are now automatically translated into the ERP system, instantly updating all related records, and outgoing files are automatically transferred back to customers and suppliers. Now there's never a need for manual data entry, so the company benefits from accurate, automatic, and timely communication across the entire supply chain.

NCA relies on EDI data to set daily schedules, forecast demand, and communicate with customers and suppliers; this new automated processing of EDI data has resulted in new levels of accuracy because the company also uses RealTime Production Monitoring by IQMS. It supports powerful, graphical scheduling screens and reports that can be used by anyone, from anywhere to assess job status, track downtime, view quality data, and more. Previously, NCA had created an infinite schedule based only on demand. Now, RealTime uses a graphical, finite schedule to assess not only machine capacity but labor capacity as well.

NCA also uses the Quality Management suite of products to control pre-production items and statistical process control (SPC); data is now communicated more quickly, without errors, and is available for review at any time.

Because all modules are built on one database, the functionalities NCA relies on—such as the EDI translator, finite scheduling, purchasing, real-time production monitoring, and quality management—work together within the system to ensure tighter control and better visibility over the company's procedures and processes, both internally and externally, resulting in a leaner supply chain.

In addition, NCA has reduced maintenance costs by more than 70% and achieved Lean objectives such as reduced cycle time, automated workflow, and the elimination of redundant processes. As a result of the smaller number of delivery errors, better quality, and streamlined communications, the company has secured additional business from its largest customer [IQMS Manufacturing, 2015].

In Chapter 12, "Manufacturing Executions Systems (MES)," we will look at a breed of software that helps support a Lean conversion process on the shop floor.

12

Manufacturing Execution Systems (MES)

Historically, Lean was exclusively used in manufacturing. Not surprisingly, manufacturing today has a little more experience than other sectors in terms of the technology enablement of Lean processes. One place we can see this clearly is in a category of software and hardware known as manufacturing execution systems (MES).

A manufacturing execution system is a control system for managing and monitoring work in process on a factory floor; it tracks manufacturing information, typically in real time, receiving data from robots, machine monitors, and employees (see Figure 12.1). In the past, manufacturing execution systems operated as self-contained systems, but today they are commonly integrated with enterprise resource planning (ERP) software suites. The goal of a manufacturing execution system is to improve productivity and reduce cycle time (i.e., the total time to produce an order). When MES is integrated with ERP software, factory managers can be proactive about ensuring the delivery of quality products in a timely, cost-effective manner.

MES can operate across multiple function areas—such as stages of the product lifecycle, resource scheduling, order execution and dispatch, production analysis, and downtime management—for overall equipment effectiveness (OEE; a measurement of equipment-related waste), quality, or materials tracking and tracing. It can also capture and record data, processes, and results of the manufacturing process, which is especially important in regulated industries such as food and beverage or pharmaceuticals, where documentation of processes, events, and actions may be required.

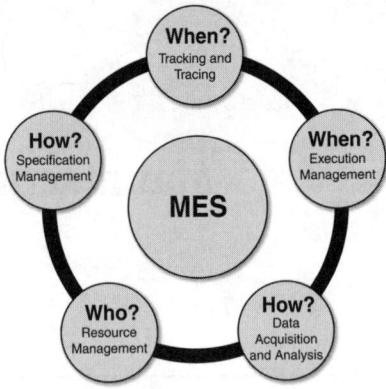

Figure 12.1 A Manufacturing Execution System (MES) Should Answer These Five Questions

The Role of MES in Today's Competitive Environment

Manufacturing execution system (MES) software can help a company to keep pace with an increasingly complex and fast-moving global environment; it can also help better align a company's complex manufacturing operations with market needs. An MES uses three key functionalities to accomplish this:

- **Operations management**—An MES can help you better coordinate the totality of production operations for both make-to-stock and make-to-order manufacturing. This includes synchronizing production activities, improving work-instruction delivery to plant personnel in complex flexible-manufacturing operations, and better managing the flow of materials from warehouses and suppliers.

- **Integration gateway**—An MES also enables integration between plant-floor and business networks. This is the key to bridging historically separate information technology (IT) and operations technology (OT) systems to create a *connected enterprise*—a unified network architecture that connects the people, processes, and technologies across an entire organization.

- **Information management**—Automating data collection in an MES can replace costly, time-consuming, and potentially mistake-prone manual data collection. Data can be displayed in dashboards and as KPIs for better decision making and consistent performance measurement. Production data can also be used for regulatory compliance and warranties, while genealogy and traceability can help limit the scope of recalls and reduce containment response times.

Manufacturing Execution Systems and Lean

Production environments face constant pressure to produce better and faster in order to remain competitive. It is a continuous struggle to eliminate unnecessary production costs; improve manufacturing, process, and business performance; increase throughput; reduce cycle times; and maintain quality. An adequate level of automation helps support these goals. Efficient manufacturing operations can be provided by a manufacturing execution system. Today, it is impossible to achieve efficient manufacturing operations without software support in many cases due to legal provisions (e.g., tracking and tracing in the food and pharmaceutical industries), high product mix, etc. Company strategies and business requirements need to be continuously adjusted to follow the latest customer requests and industry trends. Production environments are dynamic, with constantly changing products and manufacturing processes. Today's challenging business environment drives the adoption of continuous improvement (CI) initiatives such as Lean and Six Sigma in pursuit of business and operational excellence. An MES contains an abundance of information that can be used to support the process of identifying and eliminating waste. Unfortunately, a lot of companies fail to exploit the full potential of the available data. A Lean MES framework structures an approach to integrate both strategies. The result is a standard analytical environment on top of real-time data collection.

Most software implementations are done in phases. An incremental MES implementation can start with functionalities such as data collection, order tracking, material tracking, and KPI reporting. As each implementation is somewhat unique, a thorough analysis is needed. At the start, user specification requirements are set up, along with an as-is model of the current process. While the company is modeling the as-is processes, everyone is forced to question the current way of working. Problems are uncovered and inefficiencies revealed; the company is then able to create a to-be model, and this helps define the conditions for initial MES selection.

Goodness of fit and flexibility of software can be conflicting goals. Packaged systems may have best practices but may not be best suited for your organization. It is important to balance natural tendencies toward overcomplication, overautomation, and rigidity in order to get the benefits of Lean and MES systems. As processes will change over time, it is important that an MES be flexible as well.

Over time, there are three general types of changes to be aware of when implementing an MES:

- **Updates**—Small incremental (automated) software improvements may be made to fix errors and bugs or to boost performance.

- **Operational changes**—Changes may be introduced within the boundaries of a company's existing manufacturing operations management (i.e., new product introduction, installation of a new equipment, revised product flow).

- **Model changes**—More radical (and more expensive) changes may require a change of the system itself.

By integrating Lean tools such as value stream mapping (VSM) into an MES process, historical information within MES can be used to support or validate VSM analysis, and production control can be shifted from push to pull. For example, historical information within an MES can support Lean analysis by providing takt time and cycle times, leading to the selection of the pacemaker process and the calculation of kanban sizes. Production dispatching then becomes an e-kanban system, with extra process segments configured as kanban walls. Production tracking then provides the essential real-time information about the flow of kanbans. Ultimately, MES technology must be able to support change to a process, as well as control the achieved improvements [Cottyn, 2011].

As previously mentioned, an MES is often used in conjunction with an ERP system to streamline and enable actual manufacturing processes. An MES is useful for delivering standard work instructions, charging for labor, and recording quality checks.

MES, Lean, and Visual Management Systems

Visual management, when used as a Lean tool, can help regulate inventory levels and production activities. Specifically, the use of visual signals traditionally found on the shop floor provides information such as the following:

- Production line or work cell instructions that answer the questions who, what, where, when, and how for production

- Accountability and ownership for labor and support staff

- Process indicators to maintain safety, quality, on-time delivery, inventory, and cost

- Warnings of abnormal conditions (e.g., right part/wrong place, wrong part/right place, missing part)

- Production status (e.g., ahead, behind, on schedule)

A visual management system can also, through an MES, enable upstream suppliers and downstream customers when linked in the ERP system to provide better supply chain visibility. Furthermore, awareness generated through a visual management system helps enable improvement of related enterprise metrics. For example, visual work instructions help companies achieve standard work by presenting the employees with the information they need to know when they need to know it.

Some visual work instruction software only presents the information to the employees in a visual format and does not capture production information or give live visibility to the production floor. Where possible, it is important that the MES provide this information in a way that is easy to understand and allows the employees to interact with the instructions in a way that captures all the necessary production information.

Another example of visual tools in MES is the use of product flow boards. Such boards can be manual, with operators moving magnets on whiteboards, or they can be created using entry and exit bar code scans to keep electronic product status boards current.

Lean Supply Chain and Manufacturing Execution Systems Case Studies

The following sections provide some actual examples of companies that have used MES technology to achieve a competitive advantage.

Case 1: Full Sail Brewing Taps Manufacturing Intelligence to Enhance Brewing Process

Challenge

Full Sail Brewing Company, a top 25 Oregon-based brewery, produces three core varieties of craft beer, along with a number of seasonal and specialty brews.

The company realized that future production demands strain the manual mash filtration system. In addition, it recognized that efforts to increase efficiency, cut costs, and further improve quality were limited under the existing system. The mash filtration system at Full Sail required continuous manual data testing and reporting. In addition, the spent grain—a byproduct the company sold as livestock feed—contained 82% moisture content, so liquid was going out the door with the byproduct, and transporting the heavy waste was expensive. Basically, Full Sail was losing money on the transaction, paying farmers to take the spent grain.

Approach

Full Sail's goal was to upgrade its processes to improve product quality and increase filtration efficiency, capacity, and throughput. The company also wanted to minimize operator dependency by implementing automation. It decided to upgrade its traditional manual lautering process (where mash is separated into the clear liquid wort and the residual grain) to a fully automated and networked mash filtration system. The new mash filtration system Full Sail implemented leverages the PlantPAx Process Automation System from Rockwell Automation, which incorporated role-appropriate, real-time key performance indicators (i.e., manufacturing intelligence) that Full Sail can use to improve operations.

Results

Real-time data is now retrievable over variable time spans, which helps achieve optimum functionality of the system and catches discrepancies or problems that may have occurred during a batch.

With the PlantPAx system, brewing capacity has increased by 25%, and the time of each brew cycle has been cut by almost half. The visibility operators now have into their brew process allows them to optimize ongoing brews in real time. The new MES has also had the following benefits:

- Cut raw material costs by 5% annually
- Removed significantly more moisture from the spent barley grain so the byproduct is now sold at a profit
- Decreased water use by 1 million gallons annually, helping to supplement Full Sail's sustainability goals [Rockwell Automation, 2011].

Case 2: Merck Sharp & Dohme (MSD) Pharmaceutical Asian Factory Implements MES

Challenge

A new Singapore-based pharmaceuticals factory that is owned and operated by Merck Sharp & Dohme (MSD) wanted to minimize risk by getting the plant operational and then implement MES before the plant reached full capacity.

Approach

Merck considered three different MES packages. It found Ci Precision's DMS suite of products to be the most easily configurable and had the greatest scope for future expansion. Merck wanted to use Ci-DMS to control all dispensing and material addition in the factory, both automated and manual, and planned to seamlessly integrate it with the existing ERP system and automation hardware.

The system would have two supervisory stations—one on the shop floor and one for system administration. There would also be two manual dispensing stations, with a Ci-DMS Silo Dispensing module to control dosing from three "big bag" silos into an intermediate bulk container (IBC). The plan was to use bar code readers and a specially configured interface to communicate directly with the plant's ERP system and the AZO automated bulk dispensing system.

Results

The use of remote support saved around $60,000 by reducing the need for site visits. Manual dispensing and automated dispensing have both been reduced from two-person operations to one-person operations. This resulted in an annual saving of $150,000. Furthermore, the software has eliminated more than 50,000 manual transactions that were previously entered in the ERP system.

The entire dispensing operation is now paperless, resulting in a reduction in overhead. The dispensing cycle time has been reduced as there is no longer a need for manual checks to be made or for paperwork to be completed. Quality assurance has been improved significantly through the use of bar code scanners to reduce the chance of operator error [CI Precision, 2016].

Case 3: EZ-MES Automates Production Tracking System at a High-Power Laser Company

Challenge

A manufacturer of high-powered lasers offers a variety of specialized laser products that use similar production flows. During the production process, individual product data is recorded and items are serialized, requiring a complete history tracking, including subparts. To do this, the company used a combination of spreadsheets and a paper-based traveler system. The process was manual and required someone to physically search for all the parts.

As demand increased for the company's products, it became very important to be able to access accurate, real-time production status information. The company needed to know where the lasers were in the production process; inventory status of raw materials, work in process (WIP), and finished goods; who performed a specific operation and when was it done; and recent throughput measurements. The system had to be able to give real-time information about the location of all parts and products on the production floor, create travelers with bar codes that could be printed and show a list of bar coded attached parts, and attain complete traceability of parts, including when parts are assembled into other parts. The new system needed to have no significant negative impact on the current IT structure.

Approach

After comparing a number of systems, the company chose to implement the EZ-MES system. An account was set up online, and EazyWorks implemented a first production flow in EZ-MES. No training was needed, and most support during the implementation was supplied with the built-in chat feature of the EZ-MES system. No hardware was required, and no software had to be installed.

Results

Once the EZ-MES was up and running, it replaced 50% of all the paperwork work. It has helped enable growth as the company would not have been able to handle the increased volume and increased product mix without EZ-MES.

The system had a low implementation cost; the configuration of the system is entirely maintained by the client, and no training was needed. The IT department has not been burdened by hardware, software, or any additional load.

The company has been able to reduce inventory by knowing exactly what is on the production floor in real time, and it can quickly tell what raw materials to buy for each order and when to start a new batch for the same order.

Real-time information is very valuable to the laser manufacturer because it allows the company to make immediate decisions. In the past, the company had to err on the side of caution, which often resulted in too much work in progress or too many parts on order. Thanks to EZ-MES, the company has reduced inventory and WIP, improved cycle times, and increased revenue.

The company has also improved operational efficiency because the system has improved visibility for management, and the current production status is now available in real time, which enables quick and efficient decision making. The company has realized the following additional benefits:

- Improved data integrity, transparent handling of nonconformities, and less time spent manually entering data into the system
- Decreased employee turnover because the new system is easier to use

- Improved customer retention because the company can quickly respond to customer questions about specific products
- Ability to show potential customers that the system provides full traceability of all parts [EazyWorks, 2016].

The short-term, detailed scheduling of production of goods and services that is often managed by an MES can be very complex and can involve wasted capacity, firefighting, and poor customer satisfaction. The combination of Lean pen-and-pencil tools and technology can be combined to help improve and even optimize these decisions, as discussed in Chapter 13, "Short-Term Scheduling."

13

Short-Term Scheduling

A s discussed in Chapter, 12, "Manufacturing Execution Systems (MES)," if a business is capable of using Lean tools such as JIT, visual workplace, and quick changeover on the shop floor, especially with the aid of an MES, it can take full advantage of these improvements by using sophisticated planning technology. While one could argue that short-term scheduling involves planning, it is also directly tied to day-to-day manufacturing and hence its inclusion in Part IV, "Make," of this book.

Scheduling deals with the timing of operations, and the main objective is to allocate and prioritize demand (generated by either forecasts or customer orders) to available facilities in the most effective and efficient manner. The methods can range from "back of the envelope" and spreadsheets to optimization using tools such as linear programming.

Effective and efficient scheduling can give an organization a competitive advantage via the faster movement of goods through a facility and better use of assets and lower costs, additional capacity resulting from faster throughput, and improved customer service through faster and more dependable delivery to the customer. However, effective and efficient scheduling is not easy in manufacturing, as there exist the major (and sometimes conflicting) goals of short-term scheduling to minimize processing time, maximize utilization of assets, minimize work-in-process (WIP) inventory, and ultimately minimize customer waiting time.

Service systems differ from manufacturing in that there is seldom inventory (at least in the case of a pure service business, such as insurance), and scheduling is more about matching staff to variable demand. Complicating things further in service is that fact that sometimes legal or contractual issues may constrain flexible scheduling (e.g., inflexible union rules).

A manufacturing schedule decides which "job" (i.e., a customer order or forecasted demand, in the case of make-to-stock processes) can have resources allocated to it and for how long. The short-term scheduler determines the sequence in which jobs run, putting the highest-priority jobs first. This may involve making on-the-spot decisions such as interrupting a process and changing it or swapping it out for another job. The short-term scheduler must then redo the schedule without sacrificing much, if any, capacity or output. As a result, the job of short-term scheduling involves detailed knowledge of the organizations' priority rules, equipment setup and run times, job routing, scrap rates, scheduled downtimes, and so on: It's a blend of art and science.

Scheduling is dynamic, and rules need to be revised to adjust to changes, and they do not necessarily look upstream or downstream or beyond due dates. Therefore, it may be necessary to use fairly sophisticated software. To manage this, a breed of software known as *finite capacity scheduling* (*FCS*) software can overcome the disadvantages of rule-based systems by providing an interactive, computer-based graphical system. FCS software often includes rules and expert systems or simulation to allow real-time response to system changes and also allow the balancing of delivery needs and efficiency.

Short-Term Scheduling Process

At this point, it is important to note the difference between the medium-term planning models described in Chapters 7, "Master Production Scheduling," and 8, "Sales and Operations Planning (S&OP)," and the short-term, detailed scheduling models discussed in Chapter 12 and this chapter:

- **Medium-term planning**—A medium-term planning model is designed to allocate the production of the different products to the various facilities in each time period, while taking into account costs related to inventory holding and setup, transportation, and lateness. Aggregate planning looks at different product families but usually doesn't differentiate between different products within a family. It may determine the lot size for a product family at a facility. MPS, on the other hand, does schedule at the finished good SKU level but is usually in weekly planning buckets.

- **Short-term planning**—On the other hand, a short-term detailed scheduling model is usually confined to a single facility and takes into account more detailed information than does a planning model. Such models typically plan in daily (or possibly in shifts) or even hourly planning buckets and can include both interrelated independent and dependent (e.g., components or modules) scheduled production. There are usually a number of jobs, and each one has its own parameters. The jobs have to be scheduled so that one or more objectives are minimized (e.g., lateness, average completion time).

Medium-term planning and short-term scheduling models also should tie to long-term strategic models, facility location models, demand management models, and forecasting models. If there is a disconnect, service and profitability will be less than optimal.

Continuous Versus Discrete Industry Scheduling

Short-term scheduling in continuous manufacturing industries (e.g., chemical and food and beverage industries) and discrete manufacturing industries (e.g., automotive and consumer electronics) differ significantly.

Continuous manufacturing industries (see Figure 13.1) typically have main processing operations with very high changeover and fixed costs. Scheduling tools in this area can be quite sophisticated and include cyclical scheduling procedures and mixed-integer programming approaches.

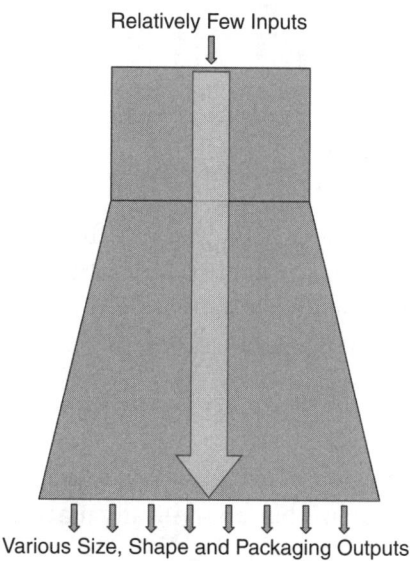

Figure 13.1 Continuous Manufacturing

Continuous industries also may have finishing operations that convert the output from the main production facilities. This may involve cutting of the material, bending, folding, and possibly painting or printing, usually with a mix of make-to-order (MTO) and make-to-stock (MTS) production strategies. Sequencing of customer orders may be important (MTO), as are forecasts and inventory targets (MTS).

Discrete manufacturing (see Figure 13.2) may involve three operations: converting such as cutting and shaping of sheet metal, main production, and assembly operations.

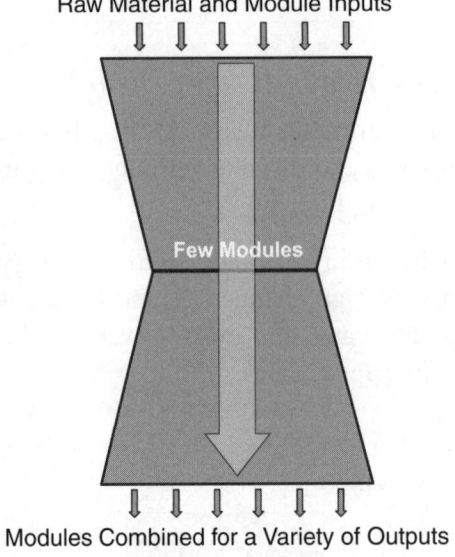

Figure 13.2 Discrete Manufacturing

The end product of converting is usually not a finished good and usually feeds a downstream operation. Main production operations require multiple different operations using different machine tools. The product and its parts may have to follow a certain route through the facility, going through various work centers. Each order has its own route through the system, quantity and processing times, and shipping date.

Assembly operations may be organized into work cells or assembly lines and usually require material handling systems but not typically machine tools.

There are some basic differences between the parameters and operating characteristics of discrete and continuous facilities:

- The planning horizon in continuous manufacturing facilities tends to be longer than the planning horizon in discrete manufacturing facilities.
- In discrete manufacturing facilities, plans and schedules may have to be changed or adjusted more often and, as a result, planning and scheduling tends to be more reactive.

- In discrete manufacturing, there may be a significant amount of mass customization and product differentiation. In continuous manufacturing, mass customization does not play a large role. The number of SKUs in discrete manufacturing tends to be significantly larger than the number of SKUs in continuous manufacturing.

As a result, the planning and scheduling issues in discrete processes can be very different from those in continuous processes [Kreipl and Pinedo, 2004].

Scheduling in Service Industries

In the service industry, the primary scheduling issue is employee scheduling at a reasonable cost to meet demand, and it varies with the time of day and the day of the week.

In general, there are three kinds of staffing problems:

- **Shift scheduling**—This type of problem involves finding the optimal crew size and assigning each member of the crew to various shifts during the day.

- **Days-on scheduling**—Some companies need to define the working and not-working days of employees in their work schedules in order to satisfy their days-off requirement.

- **Tour scheduling**—This type of problem involves both shift scheduling and days-on scheduling, and it plans working days of each employee in a planned period that is covered by the schedule. The objective of tour scheduling is to find the least costly way to staff the system at the targeted levels by determining the ideal number of employees to be assigned to each feasible work schedule.

Some application areas of staff scheduling are hospitals (especially in nurse scheduling), mail-processing organizations, trucking systems, airlines, call centers, catering, and housekeeping workers in the hospitality industry. Solution techniques include the following:

- **Manual "back of the envelope" techniques**—While not optimal, these techniques use common sense and experience.

- **Cyclical scheduling**—With this type of technique, the objective is to meet staffing requirements with the minimum number of workers. Schedules need to be smooth, and personnel need to be kept happy. Many techniques exist, from simple algorithms to complex linear programming for optimal solutions.

Lean Short-Term Scheduling

Planning and scheduling in a global supply chain requires the coordination of operations in all stages of the chain, so the models and solution techniques described above have to be integrated within a single framework. Separate models that represent successive stages in the supply chain have to exchange information and interact with one another in various ways. For example, a continuous model for one stage may have to interface with a discrete model for the next stage.

Planning and scheduling procedures in a supply chain are typically used in various phases. It is common to have a first phase with a multistage medium-term planning process (using aggregate data) and a following phase that performs a short-term detailed scheduling at each one of those stages separately. Typically, once a planning procedure has been applied, each facility can apply its own scheduling procedures. However, scheduling procedures are usually applied more frequently than planning procedures [Kreipl and Pinedo, 2004].

Advanced Planning and Scheduling (APS) Systems

Historically, manufacturers have relied heavily on the planning functionality in their ERP legacy systems. These systems are based on early 1980s concepts such as infinite capacity, time buckets, and backward scheduling. The software that drives MRP applications was primarily designed to address the needs of make-to-stock manufacturers and entailed the use of excess buffers of inventory and time in all levels of the manufacturing process.

On the other hand, traditional Lean systems tend to be manually intensive, and over time, they tend to become disconnected from a company's legacy planning systems. In addition, Lean is more difficult to implement in companies that have many SKUs, limited capacity, and unpredictable demand, which is part of the reason for the high failure rate of Lean initiatives in the United States.

Advanced planning and scheduling (*APS*) systems, can somewhat bridge the gap between ERP systems and Lean thinking. They can accurately manage time, react to changes at the operation level, and still create a schedule that quickly and accurately synchronizes multiple constraints. An APS system is a manufacturing management process by which raw materials and production capacity are optimally allocated to meet demand. APS is especially well suited to environments where simpler planning methods cannot adequately address complex trade-offs between competing priorities. Production scheduling is very difficult due to the interaction of limited capacity and the number of items/products to be manufactured.

Finite Capacity Scheduling (FCS) Systems

Specific tools of APS that are used to deal with the complexity from a short-term scheduling perspective are known as *finite capacity scheduling (FCS)* systems (see Figure 13.3). FCS is an approach to understanding how much work can be produced in a certain time period, taking into considerations limitations on different resources. The goal is to make sure that work proceeds at an even and efficient pace throughout the plant. Software applications for determining the best way to schedule work are called *decision support tools*. Finite scheduling tools are different from infinite capacity scheduling tools. Infinite scheduling tools, which are simpler, don't account for limitations on the system that occur in real time.

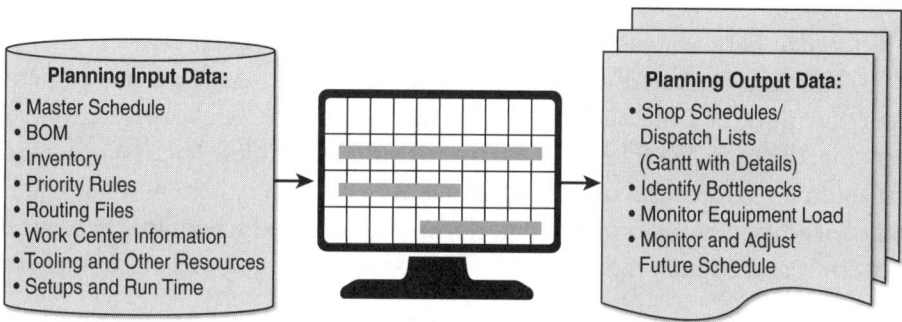

Figure 13.3 Finite Capacity Scheduling System Inputs and Outputs

The following are some of the different types of finite capacity scheduling tools:

- **Electronic scheduling board (ESB)**—ESB provides a graphical view of all jobs currently in production. If the digital board receives data from the factory floor, it can calculate performance times automatically, even if the administrator makes changes, and issue a warning if there's a bottleneck.

- **Order-based scheduling (OBS)**—The scheduler (a person or software application) prioritizes which work will be completed first by selecting only the orders that meet the plant's preset work-in-progress (WIP) criteria.

- **Constraint-based schedulers (CBS)**—Bottlenecks in the production line determine the schedule for the rest of the components in the system.

- **Discrete event simulation (DES)**—DES models random events and predicts the domino effect that one event would have on the rest of the system.

- **Genetic algorithms**—These algorithms are similar to the scientific theory of natural selection, where new schedules (children) are developed by using characteristics such as sequences of work from previous (parent) schedules.

Advanced Planning and Scheduling Systems Technology

Many of the large ERP software vendors offer APS modules. In addition, "best of breed" solutions are offered by vendors such as Asprova and Preactor.

Detailed scheduling software is an important tool for many companies and can have a major impact on the productivity of a process.

The difference between planning software and scheduling software is that planning systems are "bucketed" (monthly, weekly, daily) and don't preserve operation sequences within the time bucket. Scheduling systems are "bucketless" and preserve sequencing; they can generate dispatch lists or shop schedules. The assignment of an operation to a resource is a critical component of an FCS system and helps achieve operational efficiency and optimized performance. Detailed scheduling uses a shorter time horizon and a much more detailed process route than does a planning system.

The inputs to an FCS system are manufacturing work orders. A process route associated with each order defines the operation steps to make the product. The user can then load the orders onto individual resources, using scheduling rules, and interact with the schedule using the Gantt charts and plots that are generated. A typical output would be a dispatch list for each resource.

Advanced Planning and Scheduling Systems Technology Case Studies

The following sections provide some actual examples of companies that have used APS technology to execute effective and efficient short-term schedules.

Case 1: Auto Parts Manufacturer Chooses Asprova for Its Good User Interface Reduces Labor of Adjusting the Schedule

Challenge

An auto parts manufacturer had been using a custom-made scheduler before switching to Asprova, an APS system. Over time, as the number of products increased, the company exceeded the capability of its existing scheduling system, which caused issues such as the following:

- The inadequate data processing capability of the existing custom-made scheduler forced the company to limit the number of products and divide orders into several scheduling runs.

- Numerous typing errors that occurred when manually scheduling products left items out of the schedule.

- An inability to restrict colors at each facility led to manual schedule modifications by an experienced schedule manager.

Approach

After learning more about Asprova's functions by using a trial version of the software and attending a training seminar, the company decided to introduce Asprova. Major attractive features were the ease of understanding and modifying the schedule in Gantt chart form and the ability to restrict colors in each facility.

The company decided that the initial implementation work, including worker education, creation of the master data based on schedule managers' knowledge, and development of peripheral functions, were to be done mainly by the company itself, with the support of a consultant. Development of peripheral functions was done in parallel with the test run, but designing and altering the functions to meet the demands from the production site was not an easy task.

Results

Since the installation of Asprova, the auto part manufacturer is now able to schedule all products simultaneously and can create a complete schedule in the same amount of time it took to create one of the partial schedules before. The time required to make a working schedule has been significantly reduced because Asprova automatically creates a schedule that takes into account the color restriction of each facility, so much less adjusting work is needed.

The company has also been able to standardize most of the scheduling work by codifying schedule managers' knowledge in the form of master data that was set to Asprova at the initial installation. Furthermore, it has become much easier to explain the procedure for scheduling to new employees because of how simple it is to verify and modify the schedule using Asprova's Gantt chart [Asprova, 2016].

Case 2: Mueller Stoves Reduces the Assembly Line Stops after Preactor Deployment

Challenge

Mueller Stoves, started in 2001, produces a variety of stoves, domestic ovens, and cooktops. It had been using the concept of mini-factories pulled by kanbans from the assembly line.

Despite all the efforts of Mueller's Lean manufacturing team, the kanbans were not bringing the expected results. The main issue was that the assembly lines are extremely dynamic because of the high variability of items per day on each assembly line and the high volatility of quantities and reschedules caused by unforeseen events.

While each mini-factory was producing its own scheduling, other information was not considered, such as the availability of material (supplied by other areas or outsourced). As a result, the assembly lines often did not produce as planned because the mini-factories could not respond to changes in time. They tried to supply the components for the items scheduled on the assembly lines through daily inventory lists and urgent production requests of missing parts but often had to stop the lines due to lack of parts.

In addition, Mueller faced the following problems:

- Difficulties balancing loads across several resources
- Difficulties properly evaluating production bottlenecks
- No visibility into the consequences of unforeseen events
- No possibility of coordinating a preventive maintenance plan without sacrificing productivity
- Lack of capacity analysis to give adequate responses to the demand

Approach

As a result of the limitations Mueller faced, both the planning and production departments independently began to look for solutions. Both departments independently found Preactor and APS3 (Siemens's partner for Simatic IT Preactor products) and realized that a scheduling tool based on finite capacity, which took into account materials and resources, represented a possible solution to the challenges the company faced.

The first step of the project was to generate finite scheduling only on the stamping area in order to level the loads on the assembly lines to prevent the internal parts supply shortages that were causing lines to stop.

Results

The results were impressive and highlighted that there was also an opportunity to schedule the painting area with the same principle adopted for stamping.

The following were some of the most immediate results:

- Improved visibility, reducing production uncertainty through a scheduling horizon of 5 days and a firm schedule of 1 day
- Increased reliability of the supplier to the JIT process due to the perception on the improving factory scheduling process
- Reduction of stamping stocks from 3 days to 1.5 days of parts needed for future assemblies
- Significant reduction of semi-finished stocks
- Unplanned stoppages on the assembly lines decreased from 13 to 6 hours, a 22% reduction in WIP, and a 98% improvement in overall inventory turns [Preactor, 2015]

Even if a company has done a great job planning, sourcing, and making a product (with the aid of enabling technology), it still must deliver the product or service in an accurate, timely, and efficient manner to the customer. This is the topic of Part V, "Deliver."

PART V

Deliver

14

Distribution Requirements Planning (DRP)

D istribution requirements planning (DRP) is technique used for the efficient delivery of goods that involves determining the item quantities, location, and timing required to meet forecasted demand. The goal of DRP is to maximize product availability and reduce the costs of ordering, transporting, and holding product (see Figure 14.1).

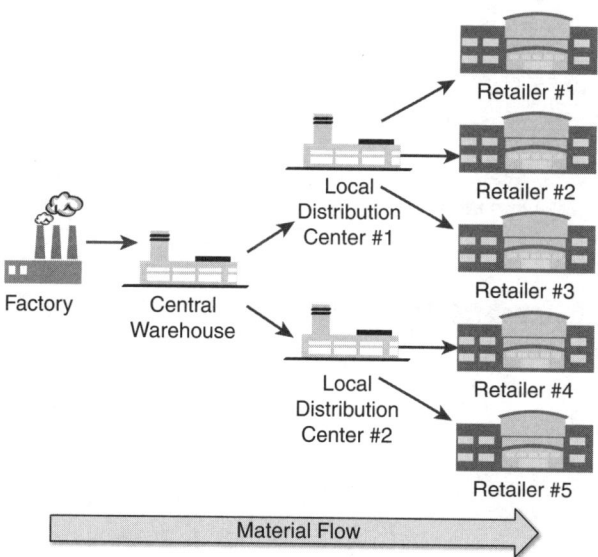

Figure 14.1 Distribution Network

DRP generates time-phased requirements for single- and multi-facility distribution organizations (i.e., where a central facility supplies regional facilities, which then supply other facilities in a tree-like structure). This process calculates inventory requirements over time and automatically generates gross and net requirements, which can be turned into inventory transfers, production work orders, and purchase orders.

The mechanics of DRP, discussed later in this chapter, are similar to those of materials requirements planning (MRP) in that DRP develops replenishment plans by evaluating information such as order size, desired safety times/service levels, on-hand inventory, scheduled receipts, and both forecasted demand and actual demand. However, in the case of DRP, replenishment requirements are for independent demand inventory (i.e., finished goods) rather than for dependent demand inventory, as in MRP.

DRP compares future forecasted demand and actual demand with available inventory (plus scheduled receipts, such as purchase orders or transfers) to predict future shortages. It also schedules planned replenishment orders (factoring in lead times) based on user-set criteria, including safety stock or safety time targets (see Figure 14.2).

DRP Planned Receipt / Planned Order Calculation

Step #1: Creating Projected Ending Inventory

Current on-hand inventory is "netted" against the next planning period's Gross Requirements. Then any Scheduled Receipts in that same period are added in to create a projected Ending Inventory for that period.

In our example, the on-hand balance of 3,300 is "netted" against the greater of the next period's Forecast (206) or Open Customer Orders (0) plus any Dependent Demand* (0) to create Gross Requirements of 206. Then, the 2,300 units of Scheduled Receipts are added in to create a projecting Ending Inventory of 5,394.

*requirements from other locations for this same item which are supplied by this location.

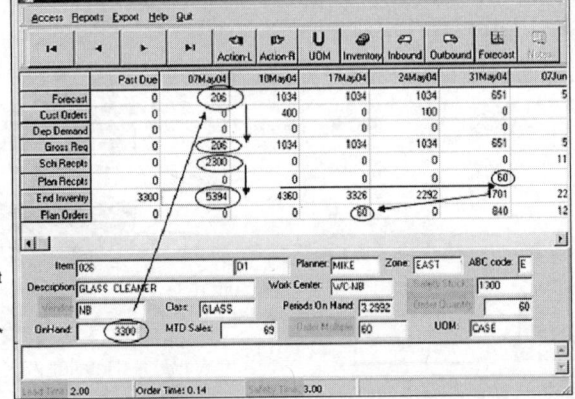

Step #2: Calculating Planned Receipts and Planned Orders

The projected Ending Inventory is netted against the Gross Requirements and Scheduled Receipts until it falls below the Safety Stock or Safety Time Quantity (whichever the user has selected to run DRP in the Configuration area).

DRP will then create a Planned Receipt when inventory is projected to fall below that level. Order (Minimum) Quantity and Order Multiples are used to calculate the exact amount of the PR. The SKU's Lead Time is then backed off from the PR date to determine when it needs to place a Planned Order at the Vendor location. If using the Summarize Demand function when initializing the system, the PO will show up as Dependent Demand for the same item at the Vendor location.

In our example, DRP creates a PR for 60 units to maintain a Safety Time of 3 periods of future Gross Requirements (1,701). The Lead Time of 2 planning periods is backed off to determine when the replenishment needs to ship from the Vendor location.

Figure 14.2 Sample DRP Screen and Description (PSI Planner for Windows, Copyright 1998–2016; Printed with Permission from Weeks Software Solutions, LLC)

DRP is "hierarchical," as the net requirements can be summarized up the supply chain to the plant level to create the item master production schedule (MPS), which can then be exploded with a bill of materials (BOM) to generate requirements for raw materials and components.

DRP is ideal for organizations that want to transition from a "push" process to a "demand pull" process, for a more efficient, Lean supply chain. However, DRP works only for companies that have the philosophy and processes in place to take advantage of a pull-type system.

Lean and DRP

Distribution inventory can either be "pushed" from the central supply down through the network or "pulled" up through the network by orders from the consumer. Pull provides the best availability for the customer because local management has control of what's available. However, it is difficult to manage distribution inventory in a pull system environment because every order is a surprise to the supplying location as demand flows up the network. Pull is characterized by the *bullwhip effect*, in which a small change in demand at the consumer end can generate large swings in demand higher up in the network and in the factory.

Push can generate the best inventory and transportation performance and can therefore result in the lowest cost. Shipments and stock levels can be centrally and globally planned. Central planning is furthest removed from actual demand, however, so the service level is likely to suffer.

DRP can theoretically deliver the efficiencies of push with the service levels of pull; however, it is dependent on the forecast and stable processes. With an accurate forecast and performance as planned (shipments arrive on time, and there are no unexpected losses, damage, etc.), DRP can deliver high service with minimal inventory. Companies use safety stock to compensate for these two sources of variability, but that reduces the effectiveness of the DRP strategy (which means more inventory and more shortages—exactly the way MRP works in the plant) [Turbide, 2016].

So, in some ways, the methods and assumptions used in DRP are opposite those in Lean thinking. DRP is all about planning ahead and optimizing, and Lean responds to the market faster and more efficiently. However, DRP settings can be adjusted to make a distribution network as lean as a company is prepared for it to be. For example, instead of pushing inventory through the network based on forecasts, a company can only use customer orders to generate deployment requirements. In today's volatile, global environment, where a company can have thousands or even tens of thousands of SKUs at many distribution points that are remote from manufacturing facilities, this is probably not the best way to go (unless you have a local manufacturing facility that is extremely flexible and agile and that ships to local customers only).

Instead, it may be best to utilize a "consumption" methodology (i.e., that looks at how customer orders consume the item forecast at each distribution center) along with scientific safety time instead of fixed safety stock targets with reorder points. This means that periods of supply inventory targets are based on variability of lead time and demand in order to catch the "peaks and valleys" of predicted demand while still considering established replenishment lead times, lot sizes, and demand and lead time variability. This can be considered a "happy medium" between pure push and pure pull when using DRP.

DRP systems are also useful as collaboration tools with customers to improve forecast accuracy, minimize retail stockouts, and improve inventory turns and productivity. We discuss this in more detail in Chapter 19, "Collaborative Supply Chain Systems."

DRP Software

Similar to many APS systems, DRP systems traditionally were add-on systems, as many DRP inputs (e.g., inventory balances, existing purchase and production work orders, forecasts) and outputs (e.g., deployment/transfer requirements, new purchase and work orders) come from or go to ERP or accounting systems. Today, many ERP vendors have added DRP as a module because the demand for it has significantly increased.

As mentioned previously, DRP software is a continuation of the MRP logic used for outbound movement of finished goods from facilities. The deployment requirements, after being summarized by item, are used to generate the item master production schedule, which then drives the MRP system through the bill of material file to create raw material and component requirements (see Figure 14.3).

DRP software is fairly similar from vendor to vendor. However, it can vary in terms of user interface and ability to collaborate and share internally and externally. In addition, like most other software today, it can be installed on a company's IT infrastructure (and may possibly be web enabled or web based as well) or, in some cases, may be offered as Software-as-a-Service (SaaS), or cloud-based, software.

MRP and DRP were pioneering technologies for the computerization of supply chain planning. In the late 1980s to mid-1990s, the APS software vendors marketed aggressively against the ERP vendors. The ERP vendors tried to downplay the importance of "best in breed" external solutions—at least until they began acquiring APS applications, including DRP, themselves. (Some ERP vendors developed the capabilities on their own rather than through acquisitions.) Eventually, many midsized to large companies implemented MRP and DRP. DRP had become more prevalent in make-to-stock (MTS) companies with extended distribution networks, such as those found in the consumer products industry.

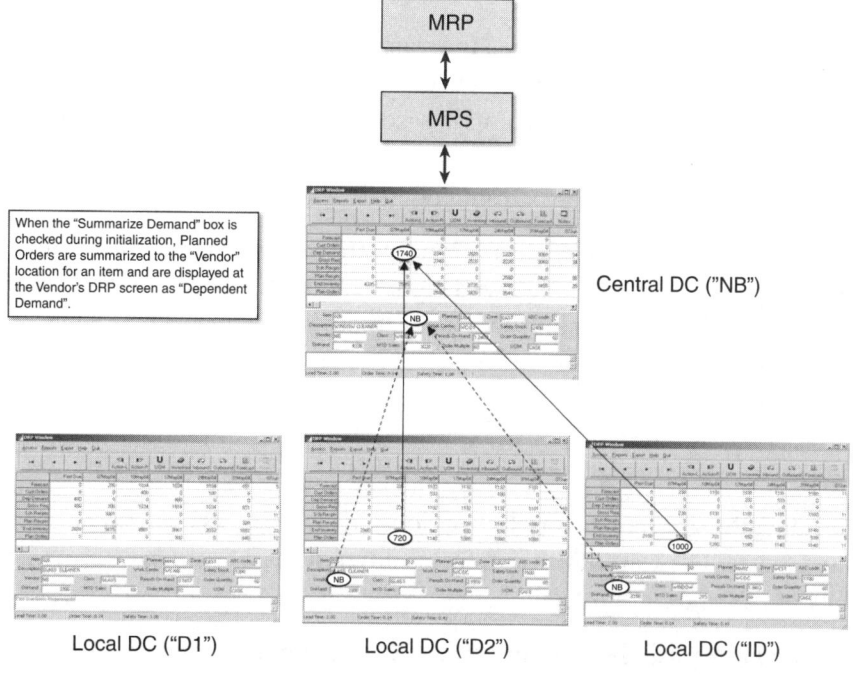

Figure 14.3 Distribution Requirements Planning in a Network (PSI Planner for Windows, Copyright 1998–2016; Printed with Permission from Weeks Software Solutions, LLC)

DRP Case Studies

The following sections provide some actual examples of companies that have used DRP or similar technology to improve the efficiency of their distribution networks.

Case 1: Ford Auto Logistics: Revving Up Service Parts Logistics Operations

Challenge

In 2011, to improve efficiency and its global service parts logistics management, Ford was in the midst of implementing a 10-year plan to revamp its service parts network, processes, and technologies. It had undergone huge changes in these three areas to create a service parts supply chain that is lean, fast, efficient, and cost-effective and to lay the foundation for the firm's global supply chain vision.

Ford's U.S. service parts operation alone includes around 250,000 part numbers (which equates to about 1.5 million SKUs), with approximately 1,500 suppliers and 4,000 dealerships.

High service levels are crucial in the auto parts business. When dealers order a fast-moving part by 4 p.m., Ford targets delivery by 10 a.m. the next day for 98% of its shipments. For bulk parts, Ford promises second-day service. In addition, like most other companies, Ford has to strike a balance between providing high service levels to customers and minimizing procurement, logistics, and inventory costs.

Prior to the changes, Ford had a traditional warehouse network model for its service parts supply chain. One hub served eight depots spread across the country that delivered parts to dealers. The hub sent parts to the depots based on a "pull" strategy, where the hub depot shipped a part when a specific facility was at risk of dropping below safety stock levels.

Approach

Ford now operates 26 smaller warehouses located closer to its customers throughout the United States. The warehouse segregates different part types by volume, size, and frequency of use in different facilities.

Ford switched to a push deployment strategy for the majority of its volume as it felt that it would have too many warehouse transfers with a pull strategy. It was able to do this by shrinking its order sizes and shipping smaller quantities more frequently using Lean techniques. As a result, the warehouse now generally receives only the parts and quantities it needs rather than getting excess parts that it puts into reserve.

One critical element of Ford's service parts supply chain change is the selection and implementation of SAP's Service Parts Management (SPM) solution. When implementing SPM in the United States, Ford first conducted a mini pilot to test the infrastructure and make sure SAP's system could talk to Ford's remaining legacy systems, and then it began launching SPM in 2010.

Results

Ford found a variety of benefits in different functional areas, including the following:

- **Inventory planning**—Ford now has the ability to optimize combined safety stock and economic order quantity values, which helps the company better determine the optimum order quantity for each part in inventory. This ends up driving down safety stock on low-volume parts.

- **DRP**—The SAP DRP system features future-dated orders and supplier shutdown logic, among other functionalities, plus it gives multiple short-date expediting options (whereas the old system had only one process for expediting parts). SAP allows the system to differentiate between suppliers' preferences on how, when, and how much inventory to expedite.

- **Parts deployment**—SAP's automatic rounding algorithm enables Ford to better marry actual demand with packaging hierarchy logic.

After Ford had fully implemented SPM in Europe only, the company realized the following benefits:

- 20% improvement in forecast accuracy

- 15% reduction in service parts inventories

- 10% reduction in obsolescence

- 10% improvement in referral cost

- 0.5% increase in local fill rates [Partridge, 2011]

Case 2: Everlast Builds a Championship Company with New Product Lines

Challenge

Everlast, founded in 1910, is an American company that is active in the design, manufacturing, licensing, and marketing of boxing, mixed martial arts, and fitness-related sporting goods equipment, clothing, footwear, and accessories.

Recently, in addition to working with retailers who are outside the usual sporting goods arena, Everlast has been expanding its marketing to specific groups in the athletics sectors, such as the Latino and women's markets. While Bed Bath & Beyond and Sharper Image handle some of the more unusual products, Everlast does the bulk of its sales through major sporting goods stores and through broad-based consumer outlets such as Walmart, Sears, Target, and Kmart.

The growing number of sales outlets is mirrored on the supply side as Everlast expands the number of overseas manufacturers it relies on. Everlast's expansion into new areas resulted in a jump from 500 SKUs to 2,000 SKUs in just five years. The company realized it needed better ways to handle forecasting and resupply.

Approach

Demand Solutions Requirements Planning (DSRP; now owned by Logility) was initially recommended to Everlast by the company's new senior vice president of manufacturing. Everlast was counting on DSRP to help deal with the increased complexity. Everlast found the application to be very user friendly. Demand Solutions had already written an interface to MAS200, the host system that Everlast uses, so the company could get a quick start.

Results

Since implementing DSRP, Everlast has benefitted in terms of speed and accuracy:

- In the past, it took 24 to 48 hours—and sometimes even a week—to roll up all the projections from sales reps. Now the company gets a starting point in half an hour.

- DSRP provides more data more quickly and in a very presentable format. Previously Everlast used spreadsheets.

- Everlast strongly believes DSRP will help keep fill rates at 99% or 100% [*DS Magazine*, 2007].

Case 3: Canadian Tire Keeps Stores Rolling with Replenishment Program

Challenge

Canadian Tire, a tire and auto parts retail chain with more than 600 tire and auto parts locations (also featuring sports and leisure and home products), realized its aging stores and out-of-date replenishment system were costing it business.

Canadian Tire operates two huge distribution centers in the Toronto suburb of Brampton, with a combined 1.6 million square feet of space. The stores are operated by independent dealers and are, in effect, the customers of the parent company. They were not happy because in-stock replenishment to the stores was below industry standards. Partially driving this was the fact that inbound service from the supplier base was around 70%, at best, in terms of on-time delivery. The goal was to get inbound on-time service levels into the 90% plus range.

Approach

Canadian Tire began to consider applying DRP to retail. It eventually selected Manugistics software (now owned by JDA), which had years of experience working with planning and scheduling issues in the consumer packaged goods and other industries. It would be the first use of core applications that were not written in-house at Canadian Tire.

The process started with a gap analysis, which led the company to change some processes so that software wouldn't have to be modified. The following quarter, Canadian Tire went live with a pilot involving three buying teams. The company took its time rolling it out over a year-long period to make sure to get it right.

Results

Previously, Canadian Tire had used in-house reorder point software to trigger replenishment requests. The heart of the new system is a rolling 26-week replenishment plan for each of the retailer's approximately 60,000 SKUs. This plan is regenerated and sent out to all suppliers weekly.

The numbers are based on actual and projected orders from the stores. The stores receive a "deal book" months in advance of each weekly flyer, detailing the items that are to be featured. These orders, as well as standard replenishment orders, are filed daily via EDI and pulled into the Manugistics system overnight. This visibility is crucial not only for promotions but also for seasonal items.

Once it was able to provide this longer view, Canadian Tire asked suppliers to reduce their purchase-order lead time to 14 days or, in a very few cases, 21 days. After one year of full operation, supplier lead times had dropped from a 46-day average to 15 days, and on-time service from suppliers was up 20%. As a result, on-time delivery to the stores had improved 1.9%. These improvements came during a period when shipment volume grew significantly (25% over 24 months). Inventory turns at the distribution centers also increased 1.7 times.

Canadian Tire also worked on other areas of execution where they could leverage the information provided by the Manugistics system. It wanted to improve transportation efficiency, inbound and outbound, and increase productivity at the distribution centers. Also, at the time, the system was driven by batch data received via EDI and did not allow for true real-time collaboration with suppliers. So Canadian Tire planned to look at using the Internet for real-time communication and collaboration with customers and suppliers [Murphy, 1999].

Whereas distribution inventory replenishment systems like DRP tell you what you need, when you need it, and where you need it, transportation management systems (TMS) focus on actually getting the product to the destination, as discussed in Chapter 15, "Transportation Management Systems (TMS)."

15

Transportation Management Systems (TMS)

The transportation systems that are used to connect your supply chain must be managed and controlled properly, with complete visibility and great communication between partners. Transportation and logistics (primarily warehouse operations) costs can account for as much as 7% to 14% of sales, depending on the industry. Transportation costs alone comprise the vast majority of this expense for most companies. Best-in-class companies have transportation and logistics-related costs in a range of 4% to 7%, depending on industry sector. So it's not hard to see how operationally and financially important transportation is to a successful business.

Not surprisingly, transportation management systems (TMS) have existed in one form or another to manage this process for quite a long while. Historically, they have been add-ons to existing ERP or legacy (i.e., homegrown) order processing or warehouse management systems. Like most other software today, TMS can be installed as resident software or web-based, and they are also available as on-demand Software-as-a-Service (SaaS).

A TMS offers benefits to an organization such as automated auditing and billing, optimized operations, and improved visibility (see Figure 15.1). Such systems typically help in planning, scheduling, and controlling an organization's transportation system, including the following functionality:

- **Planning and decision making**—A TMS helps define the most efficient transport schemes according to parameters such as transportation cost, lead time, and stops. It also includes inbound and outbound transportation mode and transportation provider selection and vehicle load and route optimization.

- **Transportation execution**—A TMS allows for the execution of a transportation plan such as carrier rate acceptance, carrier dispatching, and EDI.

- **Transport follow-up**—A TMS enables tracking of physical or administrative transportation operations, such as traceability of transport events, receipts, custom clearance, invoicing and booking documents, and transport alerts (delay, accident, etc.).

- **Measurement**—A TMS allows for cost control and key performance indicator (KPI) reporting related to transportation.

A supply chain system is made up of connecting links and nodes; the transportation system provides the links, and the facilities are the nodes. It is said that "a chain is as strong as its weakest link," and the efficient, timely management of transportation links is especially critical in today's global supply chain.

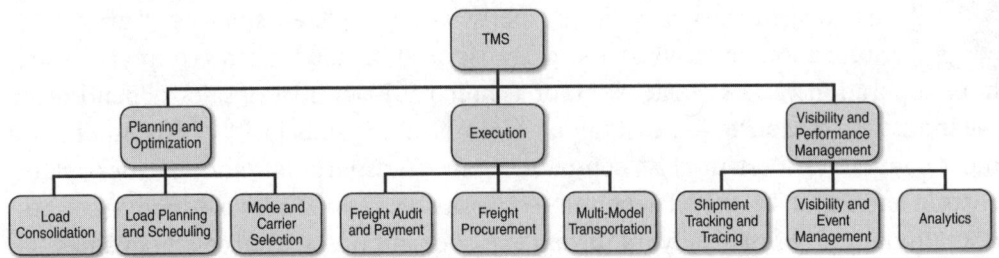

Figure 15.1 Transportation Management System (TMS) Components

A yard management system (YMS), which may be a module of a TMS (or WMS) or a standalone application, integrates warehouse operations with inbound and outbound transportation and maximizes yard and warehouse efficiency by managing the flow of all inbound and outbound goods.

A YMS enables a business to plan, execute, track, and audit loads based on critical characteristics such as shipment type, load configuration, labor requirements, and dock and warehouse capacity. A YMS is used to arrange dock appointments for receiving orders and for arranging and scheduling outbound transportation equipment and also helps in managing materials and transportation equipment in the warehouse or factory yard.

Lean and TMS

As stated earlier, a TMS helps companies move freight from origin to destination efficiently, reliably, and cost-effectively. Using Lean principles in conjunction with a TMS can result in lower transportation costs, improved customer service, reduced inventory levels, and progress toward business goals.

Therefore, it is important to focus on *total* logistics costs, not just individual costs, such as transportation, warehousing, inventory carrying costs, and so on. Furthermore, it is a good idea to apply some guiding Lean principles to your transportation system (supported by functionality commonly found in a TMS), as described by Martichenko [2016]:

1. **Develop and execute transportation strategy**—Transportation strategies should support a company's inventory strategy and customer expectations.

2. **Eliminate transportation waste**—Excess transportation can be considered pure waste. So instead, the focus for a transportation system should be as a strategic differentiator for an organization.

3. **Measure transportation performance**—Transportation providers are looked at too often from a transactional perspective—they should actually be considered strategic partners. For a Lean supply chain, transportation service providers performance should be measured and be given feedback to build a long term relationship that is stable and dependable.

4. **Understand transportation cost structure**—There is significant opportunity for transportation cost reduction in productivity costs such as trailer utilization, total miles run, equipment waiting time and adhering to core carrier routing guides, as opposed to unit costs. Having a unit cost, or carrier rate focus, can result in instability in the transportation network.

5. **Perform transportation daily event management**—Daily event management and hour-to-hour focus on waste identification and reduction can result in significant cost savings. Daily route designs, real time track and trace, real time metrics, and daily problem solving are critical as this investment and focus on process discipline is critical in Lean thinking in a transportation system.

Types of Savings and Improvements Resulting from the Use of a TMS

TMS offer a strong ROI. The primary area where a TMS can save a company money is in lowering its freight spend. Types of saving can be put into three general categories:

- Better data collection, leading to improved decision making in areas such as increased usage of core carriers and better procurement negotiations

- Process enforcement (e.g., due to ensuring that the best carriers on a lane are selected for moves)

- Optimization of processes such as loading and routing of equipment

In addition, Lean thinking leads to other non-cost-related advantages of using a TMS, including the following:

- Improved understanding of the cost to serve customers
- Reduced size of the transportation department
- Lower carbon footprint (and lower fuel costs) through optimized routing of vehicles
- Access to a greater number of carriers and increased ability to utilize capacity in the long term
- Reduced number of warehouses required
- Better tracking and compliance with transportation-related health, safety, and environmental regulation requirements

These freight savings and other benefits do not reduce service levels in any way; in fact, they are likely to significantly improve them [Banker, 2011].

Transportation Management System Technology

Because a TMS is a component of an overall supply chain management system, it is typically part of, or integrated with, an enterprise resource planning system. A TMS usually sits between an ERP or legacy order processing and a warehouse software module. Inbound (procurement) and outbound (shipping) orders typically need to be considered by the TMS to offer routing suggestions. After review by the user, the routes are then analyzed by another TMS module for selection of the best mode and low-cost carrier. The system then electronically generates electronic load tendering and sets up the shipment with the selected carrier, which also supports track and trace, freight audits and payment. This information is then fed back to ERP systems as well as possibly to a WMS system.

The following are the most common types of TMS installations:

- Traditional onsite installation and licensing (offered as a separate module by many ERP vendors)
- Remotely hosted licensing (i.e., SaaS/on-demand software)
- Onsite hosted licensing (a combination of the preceding two options)
- Hosted TMS free of licensing (the same as the second option but free, with no license requirements)

Many of the leading-edge TMS provide web and electronic interfaces to enable collaboration with transportation companies, trading partners, suppliers, and customers in some cases, are in real time. Figure 15.2 shows a route review and load optimization module from a web-based TMS.

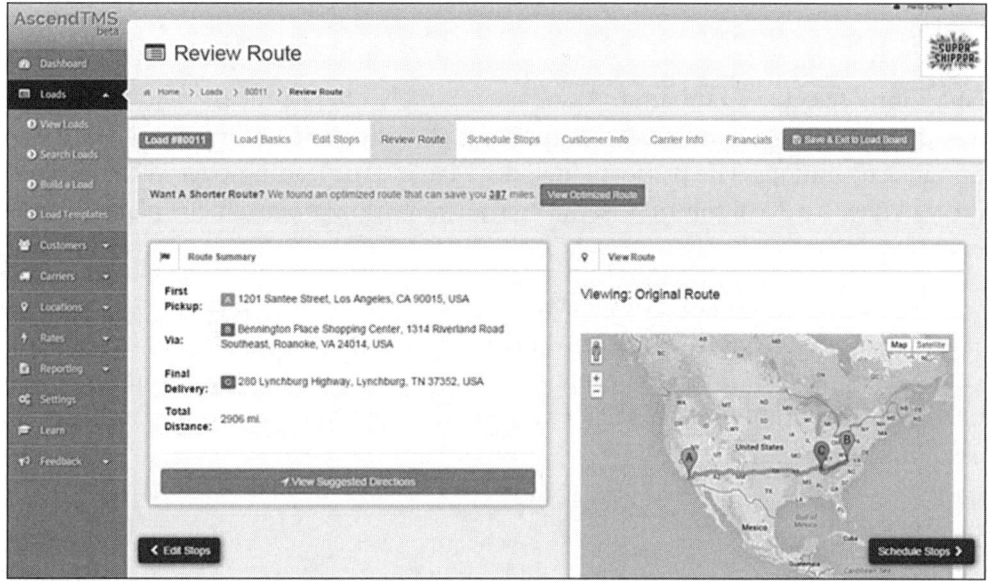

Figure 15.2 Automatic Route Review and Load Optimization Screen from Ascend TMS (Printed with Permission from InMotion Global, Inc.)

TMS Case Studies

The following section provides some actual examples of companies that have used TMS technology to improve the effectiveness and efficiency of their transportation systems.

Case #1: Papa John's Pizza Orders "Optimization Supreme" with Manhattan's Supply Chain Process Platform

Challenge

Louisville, Kentucky–based Papa John's Pizza is one of the largest pizza companies in the world, with more than 3,000 restaurants. Supply chain inventory, visibility, and accuracy issues were hurting its business. In fact, the company had to use outside storage for inventory that ended up being written off because the shelf life ran out before Papa John's could ship it to the restaurants.

In addition, Papa John's faced increases in commodity prices, fuel prices, and the minimum wage plus the usual limited-time consumer offers, which require immediate supply chain responses to temporary spikes in demand. The company needed a solution to align all the elements of its supply chain to reduce inventory levels and eliminate outside storage and inventory write-offs.

Approach

Papa John's selected Manhattan Associates's supply chain process platform, which provided the perfect solution for optimizing all replenishment, inventory, and performance operations. The platform includes Manhattan's leading-edge WMS solution integrated with a TMS solution featuring transportation procurement, planning, and execution.

Papa John's did a phased implementation of the individual solutions, starting with centralized purchasing/inbound inventory (addressed by Manhattan's replenishment and transportation procurement solutions), followed by warehouse management and then outbound delivery (with Manhattan's transportation planning and execution).

Results

The solution provided improved visibility along with reduced expenses, improved efficiency, and increased productivity in every part of the supply chain. Papa John's now knows when a purchase order was created and can track a delivery all the way to the customer. It can now manage inventory levels accurately, efficiently, and more dynamically, based on actual need, resulting in reduced overall inventory levels. It achieved a 10% to 15% reduction in freight spend after just 6 months of implementation.

Papa John's has also experienced the following transportation-related benefits:

- 10% reduction in freight costs
- 66% reduction in outside storage costs
- 83% lower inventory write-offs
- 25% improvement in vehicle cube utilization
- 11% lower mileage
- 15% increase in stops per truck
- 16% increase in tractor fill [Manhattan Associates, Inc., 2013]

Case #2: A Leading Dairy Trims 18% from Transportation Costs Using Optimizer Software

Challenge

A leading dairy required inbound transportation planning for its private fleet and for-hire carriers. It was using manual processes, moving data from various providers back and forth between an order management system and a fleet telematics tool (a device that merges telecommunications and infomatics, used heavily in fleet tracking and management). This arrangement was especially cumbersome for the company due to the high perishability and fluctuating demand for milk production. On top of that, the grades, types, and qualities of milk each have specific delivery requirements; this added complexity to the inbound transportation planning process. As a result, many hours were spent manually optimizing inbound shipments. Despite the time and effort, equipment and drivers were underutilized, which resulted in higher transportation costs.

Approach

After much searching, the dairy selected the LoadFusion Transportation Optimizer module of UltraShipTMS to control all inbound transportation planning for its private fleet of tanker vehicles. It was configured to capture the numerous inputs, such as farm pickup request (histories, schedules, frequencies, etc.) and quantity of milk produced at each farm. It was also configured to capture real-time information from clients' fleet telematics systems and the dairy's own order management system. They were now able to see that during certain peak periods, equipment utilization and capacity were strained.

With the new system, transportation planners now had the ability to review farm schedules and identify potential modifications to farm pickup request times. They were also able to balance order demand with the specific types of supplies available at multiple farms and to source products from the location with lowest transportation costs.

Results

Soon after the new system went live, transportation planners were able to determine an optimal plan that used the fewest trucks, miles, and number of drivers to pick up the maximum volume of product to meet customer demands. Route optimization saved this client 10% on transportation and an additional 8% via improved demand planning, for a total optimization savings of 18%.

In addition, the dairy was able to cut planning time by 90% through automation, reduce empty miles, and increase compliance with business rules and Department of Transportation regulations, and better enforce union work rules [UltraShipTMS, 2015].

Case #3: Miller Brands UK Uses Transwide TMS to Manage Growing Transport Volumes

Challenge

Miller Brands UK is one of the world's leading beer producers. The company's rapid growth presented new challenges to its supply chain organization, making it increasingly difficult to efficiently manage all the inbound transportation and placing a greater workload on the already taxed transportation planning organization.

Approach

Miller Brands UK looked for a TMS solution to increase efficiency, support growth, and improve customer service. Specifically, it was looking for a solution that could automate the labor-intensive transportation planning processes and provide more visibility so the company could be more proactive.

Miller Brands UK selected and implemented Transwide TMS to automate previously manual processes. Phone, fax, and email communication with carriers were replaced by a web platform for communication. Real-time transportation status updates, event notification functionality, and centralized transactional data provided increased visibility for the transportation planners and customer service teams.

Results

Processes are now centralized and standardized. An audit trail details every communication and action associated with a transportation order. As a result, Miller Brands UK has seen a 50% reduction in internal processing time for its transport operations.

The planners and customer service agents now receive email alerts when unforeseen things occur, such as incidents during the pickup or delivery process.

Thanks to integration with the company's SAP ERP system, any updates or changes to the transport are automatically updated in the respective systems.

The greater visibility of real-time information provided by Transwide TMS, has allowed Miller Brands UK to better manage direct deliveries to customers. In addition, the company has access to improve on-time delivery metrics, it has increased service levels, and it can respond more quickly and accurately to customer inquiries.

The new TMS has given Miller Brands UK a savings of $220,000 per year, resulting in a full return-on-investment (ROI) within 2 months. In the future, Miller Brands

UK is planning on using the TMS to drive its continuous improvement process. It has already reduced carrier wait times and demurrage charges but is looking to make additional adjustments in the appointment-scheduling process. It is also looking to reduce unloading "turn times" from 3 hours down to 45 minutes [Transwide TMS, 2016].

Now that we have discussed transportation management systems, which cover the links in a supply chain, it is logical to discuss the technology used in the nodes of a supply chain (i.e., warehouses and distribution centers). That is the subject of Chapter 16, "Order-Fulfillment Systems."

16

Order-Fulfillment Systems

O rder fulfillment is the process from the point of sales order to delivery of a product to the customer. In general, it refers to the way firms respond to customer orders. In many instances, the term *order fulfillment* is used to describe the act of distribution or the logistics function only. For the purposes of this chapter, we use this term to include warehouse, order, and customer relationship management and the systems used to manage these processes. The actual physical delivery of goods to the customer was covered in Chapter 15, "Transportation Management Systems (TMS)." All these processes and accompanying systems, whether referred to by these names or not, are typically linked together to help fulfill customer demand.

Warehouse Management System (WMS)

Warehouse management systems (*WMS*) are software applications used to manage the receipt, movement, and storage of materials within the "four walls" of a facility and process the related transactions necessary for receiving, put-away, picking, packing, and shipping. Early warehouse management systems provided only simple storage location functionality. Today's best-in-class systems, such as those offered by Manhattan Associates and High Jump Software, go beyond basic picking, packing, and shipping and use advanced algorithms to mathematically organize and optimize warehouse operations; they may also include tracking and routing technologies such as radio frequency identification (RFID) and voice recognition.

While many ERP vendors include WMS modules, in many cases, companies license WMS from vendors that specialize in that type of system and then integrated them with their ERP or accounting systems. They can be run as installed systems or as cloud-based, on-demand Software-as-a-Service (SaaS) systems.

Speed and accuracy are paramount in a warehouse. As mentioned earlier, many sophisticated WMS are capable of using automatic identification and data capture technology, such as bar code scanners, mobile computers, and potentially RFID, to efficiently manage and monitor the flow of products in a warehouse. Once data have been collected, they are synchronized either via batch or real-time wireless transmission to a central database, which provides a variety of reports about the status of material in the warehouse.

In warehouses where there are multiple picking locations that require fast and accurate picking, a "pick to light" or light-directed system can be used to enhance the capabilities of the employees. This type of system has lights above the racks or bins the employee will be picking from. The operator scans a bar code that is on a tote or picking container representing the customer order. Based on the order, the system requires the operator to pick an item from a specific bin. A light above the bin illuminates, showing the quantity to pick. The operator selects the item or items for the order and then presses the lighted indicator to confirm the pick. If no further lights are illuminated, the order is complete.

Voice-directed picking systems are gaining popularity. In this type of system, workers wear a headset connected to a small wearable computer, which tells the worker where to go and what to do, using verbal commands. The operators confirm their tasks by saying predefined commands and reading confirmation codes printed on locations or products throughout the warehouse. These systems are used to free the operators' hands and eyes; paper and mobile computer systems require workers to read instructions and scan bar codes or enter information manually to confirm their tasks.

Order Management Systems (OMS)

An *order management system* (*OMS*) is a computer software system used in many industries for order entry and processing. In most cases, it is part of a larger ERP, WMS, or accounting system (see Figure 16.1).

OMS applications manage processes including order entry, customer credit validation, pricing, promotions, inventory allocation, invoice generation, sales commissions, and sales history.

A distributed order management (DOM) system is different from on OMS in that it manages the assignment of orders across a network of multiple production, distribution, and/or retail locations to ensure that logistics costs and/or customer service levels are optimized.

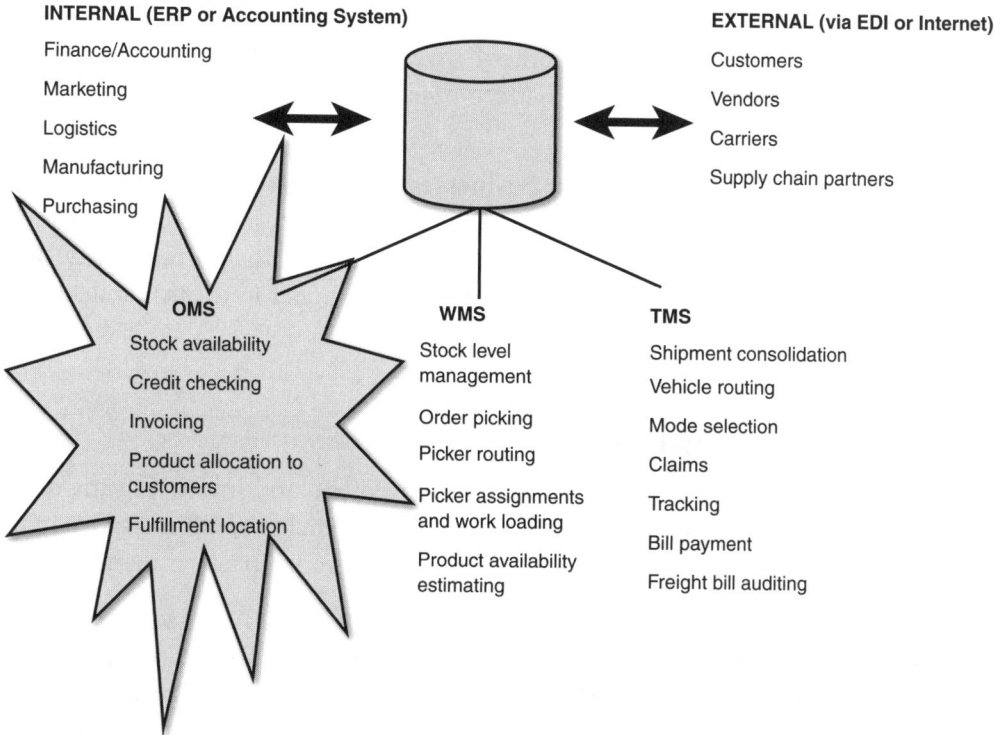

INTERNAL (ERP or Accounting System)

Finance/Accounting

Marketing

Logistics

Manufacturing

Purchasing

EXTERNAL (via EDI or Internet)

Customers

Vendors

Carriers

Supply chain partners

OMS

Stock availability

Credit checking

Invoicing

Product allocation to customers

Fulfillment location

WMS

Stock level management

Order picking

Picker routing

Picker assignments and work loading

Product availability estimating

TMS

Shipment consolidation

Vehicle routing

Mode selection

Claims

Tracking

Bill payment

Freight bill auditing

Figure 16.1 Order Management System (OMS) and Other Supply Chain Execution Systems

An OMS is usually deployed as part of an enterprise application such an ERP system as its sales engine is integrated with the organization's inventory, procurement, and financial systems.

Customer Relationship Management (CRM) Systems

The term *customer relationship management (CRM)* refers to processes, strategies, and technologies that companies use to manage and analyze customer interactions and data throughout the customer lifecycle. The goal of CRM is to improve business relationships with customers while assisting in customer retention and increasing sales revenue.

A CRM system is a software application that manages a company's interactions with current and future customers and uses technology to organize, automate, and synchronize sales, marketing, customer service, and technical support. It includes the management of business contacts, clients, contract wins, and sales leads within the sales function, using *sales force automation (SFA)* software. The four largest vendors of CRM systems are salesforce.com, Microsoft, SAP, and Oracle. There are also many other smaller CRM vendors that are popular among small- to mid-market businesses.

Perhaps the biggest benefit to most businesses when moving to a CRM system comes from having all the business data stored and accessed from a single location; before CRM systems, customer data was spread out over office productivity suite documents, email systems, mobile phone data, and even paper note cards and address book entries in various departments.

Lean and Order Fulfillment

The order-to-cash cycle is a critical Lean measurement, and improving the order fulfillment process is key to shortening this cycle. It can include administrative functions such as order management or warehouse and transportation planning and operations.

At least at a macro level, order fulfillment involves four stages: order placement, order processing, order preparation and loading, and order delivery (see Figure 16.2). If we look at these four processes, it is easy to see how they can be improved by Lean principles and tools, including standardized work, improved flow layout, visual workplace, 5S (workplace organization), value stream mapping, team building, kaizen, problem solving and error proofing, pull systems utilizing kanbans, line balancing and cellular applications, and general waste reduction.

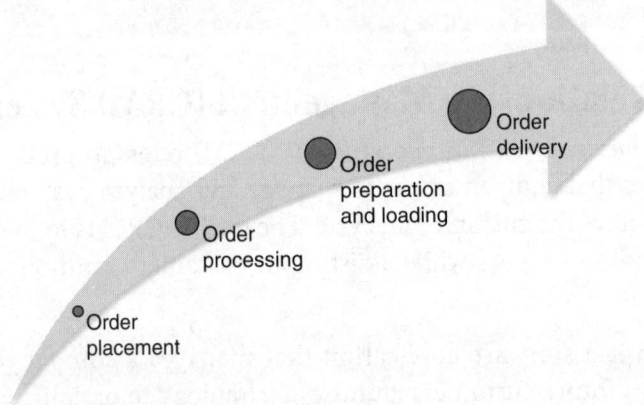

Figure 16.2 Order Fulfillment Process

Order Placement

Order placement is the series of events that occur between when a customer places or sends an order and the time the seller receives the order. A variety of order placement methods exist, including in person; by mail, telephone, fax, or electronically via EDI (electronic data interchange); and via the Internet. Electronic means such as EDI and via the Internet offer a distinct advantage in terms of timeliness and accuracy, which can significantly reduce cycle times and errors.

Order Processing

Order processing refers to the time from when the seller receives an order until an appropriate location (i.e., warehouse) is authorized to fill the order. Order processing may include the following steps:

1. Check for completeness and accuracy.
2. Check the customer's credit.
3. Enter the order into the computer system (manually or electronically).
4. Credit the salesperson (marketing department).
5. Record the transaction (accounting department).
6. Locate the nearest warehouse to the customer and advise the customer to pick the order (inventory department). Again, this can occur manually or electronically, depending on the company's technological capabilities.
7. Arrange for shipment of the order (transportation department).

A number of factors, including the following, may affect order processing time:

- **Processing priorities**—As with short-term scheduling, priorities can be first–come, first-served, shortest lead time, and so on. This varies based on an individual organization's strategy and policies.
- **Order-filling accuracy**—The more accurate the order that is received from the customer and input by customer service, the less time spent correcting it.
- **Order batching**—It may be most efficient to batch orders for picking in a warehouse. An example of a batching method is known as "wave picking," in which orders are assigned into groupings or waves and released together.
- **Lot sizing**—Full pallet orders may be processed faster than case or unit picks.

- **Shipment consolidation**—Full truckload orders are delivered faster than less-than-truckload (LTL) orders. Consolidating small orders going to the same area can not only decrease transportation costs but also speed delivery.

Value stream mapping order processing can identify significant Lean opportunities by reducing the dwell time of orders between steps such as order entry, credit checking, reconciliation, and confirmation/acknowledgement. For example, using the concept of creating a work cell for this process with cross-trained employees may cut cycle time and also reduce batch sizes and setup time.

OMS and CRM systems greatly assist in both order entry and order preparation and loading since they ensure that all current information and requirements of the customer are factored into the fulfillment process, thereby reducing the chance for errors and subsequent rework and returns.

Order Preparation and Loading

Order preparation and loading includes all activities from when an appropriate location is authorized to fill the order until goods are loaded aboard an outbound carrier. In many cases, this can be one of the best places to improve the effectiveness and efficiency of an order cycle and can account for the majority of a facility's operating cost and time. Technology such as handheld scanners, handheld computers, RFID, voice-based order picking, and pick-to-light systems can be used to speed up the process.

Order Delivery

Order delivery is the time from when a carrier picks up a shipment until it is received by the customer. It is important to closely coordinate picking and staging of orders with carrier arrival as docks and yards can get congested very easily, and charges apply when carriers are made to wait too long before loading.

Most consumer goods are delivered either from a point of production (i.e., factory or farm) in the case of larger or expedited shipments or, more typically, through one or more points of storage (i.e., manufacturer, field warehouse, wholesaler/distributor, and/or retail warehouses) to a point of sale (i.e., retail store), where the consumer buys the good to consume there or to take home.

There are many variations on this model for specific types of goods and modes of sale. Products sold via catalog or the Internet may be delivered directly from the manufacturer or field warehouse to the consumer's home. In some cases, manufacturers may have factory outlets that serve as both warehouses and retail stores.

While all processes in the supply chain in general and order management specifically are subject to measurement, order delivery is where the "rubber meets the road" and is probably one of the most critical points for success or failure in the supply chain. This is due to the fact that delivery, which in many cases may be performed by a third party, is the last point of physical contact with the customer (except in the case of returns or service). TMS, discussed in Chapter 15, assist in making this critical process more efficient.

Faster Fulfillment

We hear about companies like Amazon using technology such as kiva robots and drones to speed processing and delivery, but this goes well beyond e-commerce. Distributors, wholesalers, and "brick and mortar" retailers are moving to replenish their stores faster, keep less inventory at each retail location, and cut inventory across their entire network. For example, many retail stores now receive cartons and mixed cartons several times a week and pallets and mixed pallets less frequently, resulting in increasing their in-stock position from 90% to 97% or more while decreasing inventory by as much as 25%.

Faster fulfillment involves a trade-off between three components of an order: the order receipt, processing, and delivery cycle times (see Figure 16.3)

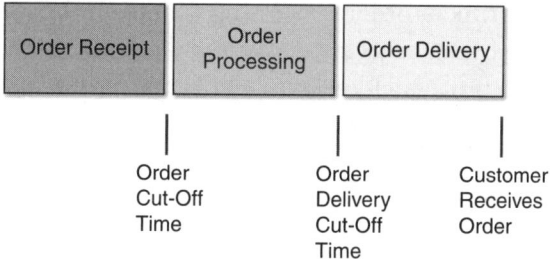

Figure 16.3 Faster Fulfillment and Trade-offs

Shortening the order processing cycle allows a company to move the order cut-off time to the right, as shown in Figure 16.3, resulting in better service to customers, and/or the ability to move the delivery cut-off time to the left, allowing access to a larger market or a reduction in the current transportation costs.

Filling e-commerce orders as a part of an omni-channel marketing and distribution strategy, while convenient for the customer, can be costly for retailers that decide to outsource their fulfillment (up to 25% of sales for some). Retailers such as Kohl's,

Walmart, Target, and Best Buy have seen reduced margins on e-commerce sales. To make this new reality work, many are rethinking their network, transportation, and distribution center operations strategies [Meller, 2015].

Order Fulfillment Case Studies

The following sections provide some actual examples of companies that have used technology to transform their order fulfillment process.

Case 1: Regional Pharmaceutical Distributor Requires Next Day Delivery

Challenge

Customers of a regional pharmaceutical distributor with a single distribution center located in the central United States require next-day delivery. The distributor had an order cut-off time of 7 p.m., but most hospitals require that their orders arrive by 6 a.m. the next day. This short transportation window gave the distributor a range, or market size, of approximately 350 miles around its distribution center.

Approach

The company carefully analyzed its situation and decided to invest in distribution center automation to shrink its order-processing window by implementing two different automated solutions: one for its fast movers and another for its slow movers. This allowed the company to process a large number of small orders at a low operating cost.

The company also decided to split the distribution center into zones rather than have a single tote travel the length of the building. Items in each zone were picked at the same time, and the totes were combined into a complete order at the end of the process.

The company chose to use the processing time reduction from these changes (a two and a half hour savings) by making the order cut-off time one hour later, which put the company more in line with its competition.

Results

The new automation technology in the company's distribution center made it possible to process more than 2,400 orders per hour, which was a large improvement over the previous manual picking process.

The combination of changes reduced the order processing window by two and a half hours, from four hours down to one and a half hours. This allowed the company to increase its transportation window by 90 minutes, which meant it could increase the size of the market it served by 75 miles (nearly 50%). This expansion eventually brought the company additional customers and increased revenue by over $100 million per year [Meller, 2015].

Case 2: Whirlpool Spins Optimized Supply Chain with Help from Manhattan Associates

Challenge

Whirlpool Corporation is the world's leading manufacturer and marketer of major home appliances, with more than 40 manufacturing facilities sourced from approximately 7,000 different suppliers. It typically has from 2.2 to 2.5 million units in inventory, delivered to about 30,000 retailers worldwide through a distribution network of 15 factory distribution centers, 10 regional distribution centers, and 85 local distribution centers.

During its integration with Maytag, Whirlpool reduced its number of major manufacturing facilities from 47 down to 25. Based on this reduction, Whirlpool felt that it needed a supply chain with the flexibility, scalability, and agility for the various requirements across multiple channels.

Furthermore, Whirlpool realized that it would first have to address these challenges:

- With a new standard of getting product to customers within 48 to 72 hours, the company realized that it had to move to a segmented inventory strategy supported by leaner processes.

- As a U.S.-based manufacturer, Whirlpool needed to lower production and distribution costs as it couldn't obtain the same labor savings as its competitors had seen by moving to overseas manufacturing.

Approach

After the consolidation, Whirlpool needed to restructure its warehouses and deploy the proper picking technologies to cut down on the number of miles workers traveled inside the facilities. The company also wanted to standardize processes and technologies across the entire network. So it had Manhattan Associates conduct a detailed needs assessment in order to understand the problems and challenges involved with the integration.

Results

Utilizing a combination of innovative solutions from Manhattan Associates, including warehouse management, slotting optimization, and labor management, Whirlpool has seen the following results:

- Eliminated 40 million travel miles between facilities, reducing fuel costs and environmental impact
- Reduced the time associated with the pick process by 50%, thereby speeding pick rates
- Cut damage to products by 50%
- Improved warehouse uptime to more than 90%, moving closer to the established goal of 99.9%
- Seamlessly integrated two competing supply chains (i.e., Whirlpool and Maytag) without customer or trading partner interruptions
- Built a more exact, real-time view of inventory inside the warehouses, which increased inventory accuracy
- Gained warehouse efficiencies through integration of pick-pack logic and use of a process with more exact inventory counts
- Improved order-to-delivery time [Manhattan Associates, Inc., 2016]

Case 3: TAGG Logistics Combines Cadence WMS and ADSI Ship-IT for Real-Time Supply Chain Execution System

Challenge

TAGG Logistics is an order fulfillment and third-party logistics (3PL) provider operating out of St. Louis, Missouri, and Reno, Nevada. It has a client base of manufacturers, wholesalers, and retailers that ship to their customers throughout North America and the rest of the world. Founded in 2006, TAGG was quickly outgrowing its legacy warehouse system.

Approach

TAGG required a new warehouse management system (WMS) and transportation management system (TMS) to do the following:

- Operate in a high-volume and diverse order fulfillment environment
- Enable the company to streamline and accelerate its shipment processing speeds and picking processes for both domestic and export shipments

- Provide flexible product and shipment tracking
- Provide least-cost routing for small parcel and LTL shipments
- Automatically capture order handling and shipping details for billing

TAGG reviewed several WMS solutions before choosing the Cadence WMS by Cadre Technologies. It chose the Cadence WMS because it was designed for the 3PL industry, provided flexibility, and included an integrated billing solution.

Next, the company replaced its TMS with ADSI's Ship-IT solution, which is tightly integrated with Cadence and offers a single platform to manage all small parcel, LTL, USPS, and regional carrier shipping.

Results

TAGG's sales have increased 53% since 2007. At the same time, the company has achieve success in the following areas:

- **Faster pick rates**—TAGG can now configure how it groups orders in waves, increasing the efficiency of its order processors.

- **Improved inventory tracking**—TAGG can now better manage cycle counts, track products in the warehouse, and manage inventory. Inventory accuracy is now over 99.9%.

- **Customer reporting**—The company can now customize and provide information to its clients.

- **Integrated billing**—TAGG has created a highly automated billing process that captures costs for all the value-added services the company provides.

- **Streamlined shipment processing**—TAGG now has a single system, which allows the company to perform least-cost routing for shipments, produce carrier labels, and upload the actual shipping costs back to the billing system.

- **Customer-specific shipping programs**—The company can automatically send customer-specific shipping rules to Ship-IT.

- **Paperless international shipping**—Staff no longer have to interrupt their workflow to manually process international shipments. Now these shipments are automatically processed in seconds [Cadre Technologies, 2016].

It would be nice if everything that shipped through an order fulfillment system to the customer never came back. Unfortunately, that's not always the case, and most companies have to deal with reverse logistics to some degree; that is the topic of Chapter 17, "Reverse Logistics Systems."

PART VI

Return

17

Reverse Logistics Systems

everse logistics is a process that until recent times hasn't been put under the microscope. Really examining reverse logistics can help companies significantly reduce waste and improve profits.

As the name implies, it is the reverse of what has been described so far in terms of plan, source, make, and deliver. Reverse logistics can be defined as the process of planning, implementing, and controlling the efficient flow of recyclable and reusable materials, returns, and reworks from the point of consumption for the purpose of repair, remanufacturing, redistribution, or disposal.

In addition, with today's environmental concerns, organizations need to try to integrate environment thinking into the entire supply chain process—forward and reverse. This includes product design, material sourcing and selection, manufacturing processes, delivery of the final product to consumers, and "end-of-life" management of the product after its useful life.

In a perfect world, of course, there would be little need for much of the material handled in reverse logistics. However, with the average manufacturer spending 9% to 15% of total revenue on returns, it is clearly a major source of waste in the supply chain. There are a variety of reasons for the reverse logistics process, including the following:

- Processing returned merchandise, including damaged, seasonal, restock, salvage, recall, or excess inventory
- Tackling green initiatives such as recycling packaging materials/containers
- Reconditioning, refurbishing, and remanufacturing returned product
- Disposing of obsolete inventory
- Recovering hazardous materials and disposing of electronic waste

Ultimately, a company's reverse logistics network can be used for a variety of purposes, such as refilling, repairs, refurbishing, remanufacturing, and so on, depending on the nature of the product, unit value, sales volume, and distribution channels.

Lean Reverse Logistics

There are five general steps to the product returns process, regardless of the industry:

1. **Receive**—Product returns are received at a centralized location, usually a warehouse or distribution center (usually after being gathered from retail locations or returned by the end user). In many cases, this step in this process involves providing a return acknowledgment.

2. **Sort and stage**—In this stage, returned products are received and sorted for further staging in the returns process.

3. **Process**—Returned products are sub-sorted into items, based on their stock-keeping unit numbers. They can then be returned to inventory. If they are vendor returns, they are sorted by vendor.

4. **Analyze**—The value and subsequent status of the returned item is determined by trained employees.

5. **Process**—Returns in good condition, such as back-to-stock or back-to-store items, are returned to inventory. If the items require repair, refurbishment, or repackaging, then diagnostics, repairs, and assembly/disassembly operations are performed as needed. Items that have been repackaged, repaired, refurbished, or remanufactured are usually shipped to secondary markets.

How well these processes are managed can go a long way toward minimizing waste.

Elements Key to a Lean Reverse Logistics Process

A research team at the Reverse Logistics Executive Council [Rogers and Tibben-Lembke, 1998] identified key reverse logistics management elements and examined the return flow of product from a retailer back through the supply chain toward its original source or to some other disposition (see Figure 17.1). The elements are described in the following sections. Depending on how they are handled, these elements can either positively or negatively impact a company's profitability.

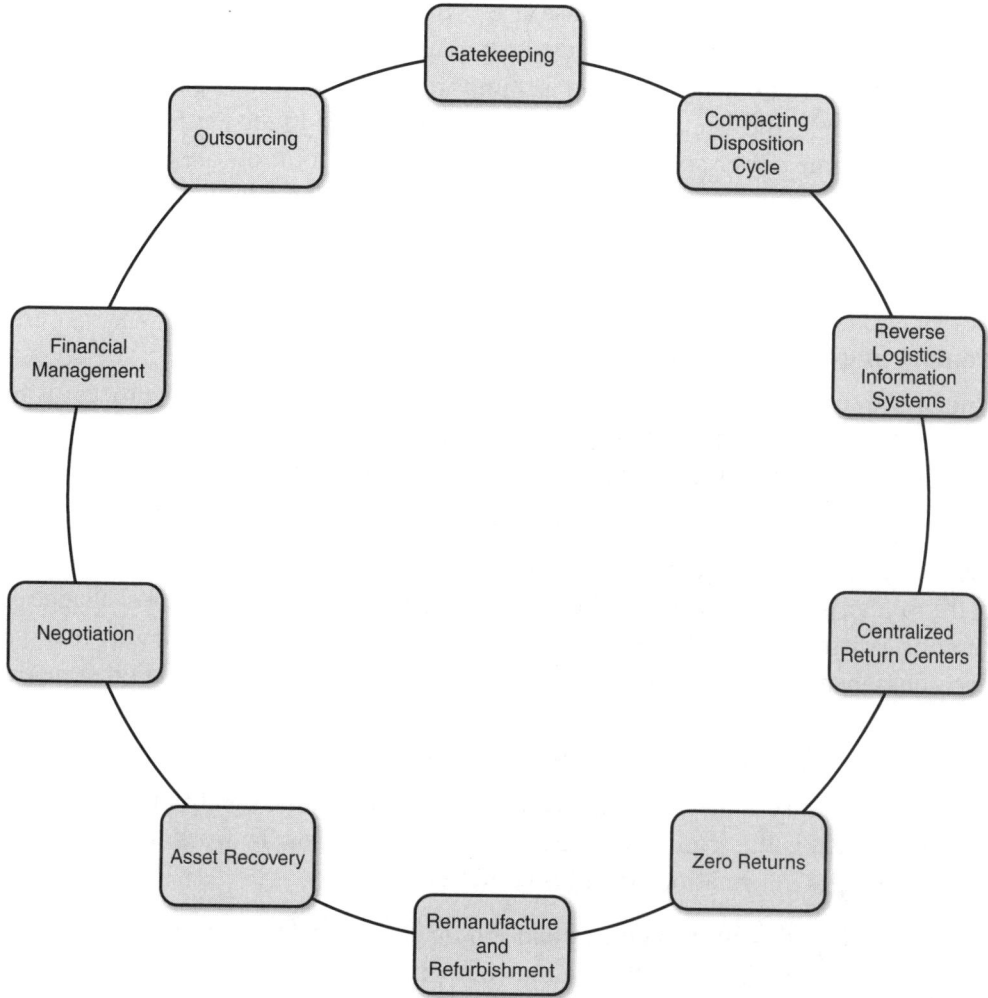

Figure 17.1 Key Reverse Logistics Management Elements

Gatekeeping

Gatekeeping involves the screening of defective or unwarranted returned merchandise at the beginning of the reverse logistics process. It is the first critical factor in ensuring that the entire reverse flow is both manageable and profitable. In the past, companies have put most resources into the forward logistics process and have given very little time and effort to the reverse process.

While liberal return policies, like those at L.L. Bean, Walmart, and Target, may draw customers, they can also encourage customer abuse, such as the return of items lightly used for an event or one occasion. So it is important to have a solid gatekeeping process. For example, the electronic gaming company Nintendo rebates retailers if they register the game player sold to at the consumer at the point of sale. By doing this, Nintendo and retailers can determine whether a product is under warranty and also if it is being returned inside the allowed time window. The impact from this system on the company's bottom line has been substantial: an 80% drop in return rates.

Compacting the Distribution Cycle Time

One of the major goals once an item has entered the reverse logistics process is to reduce the amount of time to figure out what to do with returned product. This includes return product decisions, movement, and processing.

It is important to know what to do with returned goods before they arrive. Often when material comes back in to a distribution center, it is not clear whether the items are defective, can be reused or refurbished, or need to be sent to a landfill. The challenge of running a distribution system in reverse is difficult; employees have difficulty making decisions when the decision rules are not clearly stated and exceptions are often made.

Reverse Logistics Information Technology Systems

One of the most serious problems that companies face in the execution of a reverse logistics system is the scarcity of good information systems. To work well, a flexible reverse logistics information system is required.

The reverse logistics information system should create a database at the store level so that the retailer can begin tracking a returned product and follow it all the way back through the supply chain.

The information system should also include detailed information programs about important reverse logistics measurements, such as returns rates, recovery rates, and returns inventory turnover.

Useful tools such as radio frequency identification (RFID) are helpful. New innovations such as two-dimensional bar code and RFID license plates may soon be in use extensively.

The following are some of the benefits of using reverse logistics technology:

- Goods can be tracked as they travel through the returns process. Poor product visibility in the reverse supply chain creates inefficiencies and allows for costly mistakes such as the misplacement of a returned item.

- Government environmental regulations around parts and product disposal will continue to tighten, especially in the area of recycling and disposal of e-waste. Reverse logistics technology can help with compliance.

- Reverse logistics processing costs can be reduced by implementing automation that eliminates manual processes.

- It is possible to manage and improve inventory utilization by automating directed workflow processes such as return to stock or return to supplier in the exchange and repair processes.

- Improved customer satisfaction by allowing customers to return, exchange, or repair products anywhere, no matter what channel was originally used to purchase the product.

- Linking multiple exchange or repair requests to their original sales orders enables complete order lifecycle tracking.

- Integration of manufacturers, trading partners, logistics providers, and customers enables efficient customer service.

Manufacturers, wholesalers/distributors, and retailers can all benefit from reverse logistics systems. U.S. companies spend more than $100 billion per year processing returns, and there is increased pressure from customers beginning to expect businesses to take back products regardless of the reason. In fact, research indicates that customers will stop shopping with retailers and manufacturers if the returns process is a hassle.

If reverse logistics software is fully integrated with a company's (and trading partner's) related processes and integrated with other systems such as CRM, the company can significantly improve margin recovery by increasing the efficiency of returns processing, optimizing dispositions to enable maximum recoveries, and providing timely and accurate compliance and reconciliation information through data accessibility and management tools.

Centralized Return Centers (CRC)

Centralized return centers (CRCs) can offer many benefits to an organization, including the following:

- Consistency in disposition decisions and minimization of errors
- Space-saving advantages for retailers who want to dedicate as much of the shop floor as possible to salable merchandise
- Reductions in labor costs due to the specialization of CRC employees, who can typically handle returns more efficiently than can retail clerks
- Transportation cost reductions as empty truckloads returning from store deliveries are used to pick up return merchandise
- A convenient selling tool for the easy disposition of returned items, which can be an appealing service to retailers and may be a deal-maker for obtaining or retaining customers
- Faster disposition times, which allow the company to obtain higher credits and refunds, as items stay idle for smaller periods of time and thus lose less value
- More ease in identifying trends in returns, which is an advantage to manufacturers who can detect and fix quality problems sooner than if these returns were handled entirely by customer service personnel

Zero Returns

A company may have a program that does not accept returns from its customers but instead gives the retailer an allowable return rate and proposes guidelines for the proper disposition of the items. Such policies are usually accompanied by discounts for the retailer. This type of policy passes the returns responsibility on to the retailer, while reducing costs for the manufacturer or distributor. The drawback is that the manufacturer loses some control over its merchandise.

Remanufacture and Refurbishment

The advantage of remanufacturing and refurbishment is that using reworked parts results in cost savings.

There are five categories of remanufacture and refurbishment:

- Make the product reusable for its intended purpose:
  1. Repair
  2. Refurbishing

3. Remanufacturing

- Retrieve reusable parts from old or broken products:

4. Cannibalization

- Reuse parts of products for different purpose:

5. Recycling

Asset Recovery

Asset recovery is the classification and disposition of returned goods; surplus, obsolete, scrap, waste, and excess material products; and other assets. It tries to maximize returns to the owner while minimizing costs and liabilities for the dispositions.

The objective of asset recovery is to recover as much of the economic (and ecological) value as possible, thus reducing the final quantities of waste. This can be a good cash-generating opportunity for companies that can sell these goods that would be otherwise end up in landfills.

Negotiation

Negotiation is a key element for all parties of the reverse logistics process. Because of the inherent lack of expertise on product returns, negotiations usually are informal and approached without formal pricing guidelines. Firms often do not maximize the residual value of returned product.

Financial Management

Financial management is one of the most difficult parts of reverse logistics and also one of the most important. Returns are sometimes charged against sales. Sales department personnel may tend to fight returns and delay them as much as possible. Accounts receivables are also impacted by returns.

Outsourcing

As mentioned previously, reverse logistics is usually not a core competency of a firm. In many cases, it may make more sense for a firm to outsource its reverse logistics functions than to keep them in-house.

Lean and Reverse Logistics Technology Cases

Technology can improve the reverse logistics process a variety of ways, as shown in the following cases.

Case 1: RTL™ Adds Up to Many Happy Returns for Philips Consumer Lifestyle

Challenge

Product returns cost Philips Consumer Lifestyle (part of Philips New Zealand Ltd.) hundreds of thousands of dollars each year just for the logistics. In addition, Philips faces the costs of any subsequent repairs as well as the margin loss that results from scrapping or selling products as seconds.

Approach

Although Philips had a very well-developed goods return authorization (GRA) policy, not everyone involved in the reverse logistics chain followed it. So the company started to look to reduce costs and increase efficiencies in its returns process. After a careful search, it went with RTL (Round Trip Logistics), which has an innovative web-based GRA solution.

Results

Since implementing RTL, Philips has increased customer satisfaction, reduced call center costs, and decreased staff training costs. In addition, the company now has a standardized GRA process with minimal associated IT costs.

In fact, telephone and email GRA requests to Philips's internal call center have almost totally disappeared as customers can now obtain online authorization for returns as well as arrange for transportation on RTL. Philips estimates that it saves over $100,000 per year.

The system prompts the retail returns clerk for all the pertinent information and doesn't allow a clerk to proceed until all fields are filled in correctly. As a result, the accuracy of the company's GRA forms is now approaching 100%.

Philips customers are happy because the decision making process is immediate, and any credit due is processed and applied to their accounts much faster than in the past. Philips staff are happier because RTL reduces the time they have to spend on returns activities, which means more time being able to sell. RTL has also lowered Philips's costs and increased the efficiency of the return process.

Finally, the system offers online training for returns clerks and easily interfaces with a company's ERP or accounting systems [RTL, 2016].

Case 2: Nokia—Deployment of Global Service Strategy
Challenge

Historically, the after-service repair landscape of many original equipment manufacturers (OEMs) has been primarily nationally focused. This was certainly the case for Nokia, the Finnish multinational communications and information technology company. Local organizations were typically in charge of the contacts and contracts with local repair vendors, which in most cases offered a software solution with the capability to register and manage the reverse logistics and repair cycle. The repair vendor–managed software offered a local solution that facilitated order entry, transportation management from retailers or even end consumers to the repair center and vice versa, repair and swap solutions, track and trace, and local contact center support.

Approach

Nokia decided to move away from the local infrastructure and replace it with a global model. It decided to invest in its own global software solution to get full control and visibility over all of its repair flows. By concentrating these flows to fewer repair vendors focused purely on repair, Nokia hoped to cut costs.

Nokia partnered with ReturnPool, a global IT solutions provider with expertise in the areas of business process optimization, rapid IT development, reverse logistics, and project management.

Results

In the new global process, Nokia connected its global system and the various parties involved in the reverse supply chain, took control over the reverse logistics flows, and created the ability to reroute repair volumes to different repair vendors when applicable for cost or quality reasons. In some geographic areas, logistics hubs were established to swap items when needed to meet the repair service-level agreements (SLAs) promised to customers. In the new process, the repair SLAs promised to customers could be managed separately from the SLAs agreed upon with the repair vendors if necessary.

The implementation of a global software solution wasn't as simple as just a rolling it out. It required a change management approach that considered specific local requirements when needed while maintaining and safeguarding the global focus.

ReturnPool has positioned itself as an important partner to Nokia for the past decade, assisting in the implementation of the global solution in more than 30 countries. ReturnPool has used a hands-on and global approach, making necessary adjustments to the software system as necessary in the ever-changing service landscape. Tools used include detailed process mapping, hands-on training, and onsite implementation [Return Pool, 2012].

Case 3: Return Central—Reconstructed Network Yields Big Efficiencies, Faster Processes for National Home Improvement Retailer

Challenge

A major national retailer of home improvement and construction products and services with more than 2,000 stores nationwide had an inefficient reverse logistics process in which products returned to vendors were controlled on an ad hoc basis by individual stores and regions, creating major disruptions in the reverse supply chain. Furthermore, a significant number of do-it-yourself, non-conveyable commodities created issues beyond the normal returns process that contributed to an even bigger waste stream.

Approach

GENCO, a major third-party logistics (3PL) company, developed a solution for the retailer that addressed five key areas: labor, salvage and liquidation of goods, the transportation network, visibility and management of return lanes, and recycling.

Significant resources were required to handle the retailer's returns, and this was keeping the company from its primary function of interfacing with customers to increase customer service and sales. By implementing standardization through GENCO's proprietary R-Log software, automation-enabled teammates were redirected to customer service and sales.

With R-Log, the retailer can track the movement of goods through return channels in real time. R-Log also enabled the immediate application of vendor credits to the retailer's accounts payable system. In addition to removing the burden of returns management from customer service teammates, R-Log improved tracking and visibility for returned inventory.

Results

This case illustrates the synergistic power of Lean supply chain techniques when combined with technology as GENCO implemented Lean and kaizen events throughout the returns network to create a culture of continuous improvement. To support all these strategies, GENCO also designed and built a custom reverse logistics network for the retailer. Ultimately, a regional network of three U.S. locations was executed to facilitate returns management.

The retailer's new centralized return system has been very successful. R-Log has delivered speed, efficiency, and automation throughout the complete returns process. The retailer has also had a significant reduction in the size of the workforce in returns management, and the remaining employees have been freed up to focus on other important areas of the business, such as improving the customer experience. Furthermore, the implementation of Lean processes in the three facilities has standardized process flows, improved productivity, eliminated wasteful steps, and streamlined sorting and palletizing.

The new process has reduced the losses the retailer was experiencing before engaging GENCO and its reverse logistics technology, significantly increasing recovery and revenue [GENCO, 2016].

Up to this point in the book, we have covered the use of technology in the five major supply chain processes covered by the SCOR model (i.e., plan, source, make, deliver, and return). In 2012, a sixth process, enable, was added to the SCOR model to address supporting the original five process steps with best practices. As this book has focused on using technology to enable a Lean supply chain, we will take some leeway here in terms of what was meant by this new step and focus in Chapter 18, "Measurements, Metrics, and Analytics"—specifically, on measurements and analytics used to make sure that people, processes, and technologies are aligned to *enable* a Lean supply chain.

PART VII

Enable

18

Measurements, Metrics, and Analytics

U p to now we have discussed the fundamental supply chain functions of the SCOR model (plan, source, make, deliver, and return) in terms of how technology can enable and enhance leanness. However, we still need to discuss the effectiveness of these processes and their accompanying technologies. That is where the topic of measurement and analysis comes into play. According to a Bain & Company survey of 300 global companies, "68% of managers think they have failed to optimize their supply chain savings. The ones who do, Walmart, Ford Motor, Dell Computer, all quantify performance indicators for their supply chains by setting targets that push them toward best-in-class status" [Bain & Company, accessed 2016].

Measurement and Analysis Process

As we know, an assortment of trade-offs exist in a supply chain (e.g., cost versus service) and must be counterbalanced against each other in order for a company to be successful in the long term.

It is important to fit your supply chain performance measures to your company's mission and strategy, keeping in mind that performance measures can affect the behavior of managers and employees. It is also vital to target and measure supply chain performance in order to meet customer expectations, improve supply chain capability, improve asset performance, motivate the workforce, and provide stakeholders with a satisfactory return on their investment. Using key performance indicators (KPIs) is a common way to achieve this and to see how effectively a company is achieving key business objectives. KPIs can be used at multiple levels of an organization to evaluate success at reaching targets.

Once you have established what KPIs to measure for your organization, you need to determine how to gauge yourself against them. This process, known as *benchmarking*, involves comparing one's business processes and performance metrics to industry bests or best practices from other companies.

Technology today makes it much easier to gather and analyze data through the use of exception reporting and real-time dashboards, for example. However, there is a lot more data available, making it all the more important to measure only the right things and to avoid wasted effort; otherwise, you may be subject to the dreaded "analysis paralysis." The results of analysis should be used effectively.

What and Where to Measure

While many methods are used to measure effectiveness today (e.g., the SCOR model, balanced scorecard, and activity-based costing), each having its own specifics, it is important to include the following general measurement categories in any determination of what and when to measure:

- **Time**—It is important to consider on-time delivery and receipt, order cycle time, and variability and response time.

- **Quality**—You need to measure customer satisfaction, processing, and fulfillment accuracy, including on-time, complete, and damage-free order delivery, as well as accurate invoicing. It is also important to consider planning (including forecasting) and scheduling accuracy.

- **Cost**—This category includes financial measurements such as inventory turns, order-to-cash cycle time, total delivered costs broken up by cost of goods, transportation, carrying, and material handling costs.

Using the SCOR Model to Measure and Control

We have used the SCOR model (www.apics.org) throughout this book to show where technology can be used in the supply chain. You can also integrate this model with your metrics as they relate to Lean. In fact, the model contains more than 200 process elements, 550 metrics, and 500 best practices that are organized around the six primary management processes plan, source, make, deliver, return, and enable.

SCOR metrics are organized in a hierarchical structure. Level 1 metrics, at the most aggregated level, are typically used by top decision makers to measure the performance of a company's overall supply chain. Level 2 metrics are primary, high-level measures that may cross multiple SCOR processes.

All SCOR metrics have five key strategic performance attributes, or groups of metrics used to express a strategy. While an attribute itself cannot be measured, it can be used to set strategic direction. The five performance attributes, all of which can be related to various forms of waste, are reliability, responsiveness, flexibility, cost, and asset management. Table 18.1 shows some examples of metrics, and the following sections discuss further considerations.

Table 18.1 SCOR Model Strategic Metrics

Performance Attribute	Sample Metric	Calculation
Supply chain reliability	Perfect order fulfillment	Total perfect orders/Total number of orders
Supply chain responsiveness	Average order fulfillment time	Sum of actual cycle times for all orders delivered/Total number of orders delivered
Supply chain agility	Upside supply chain flexibility	Time required to achieve an unplanned 20% increase in delivered quantities
Supply chain costs	Supply chain management costs	Cost to plan + Cost to source + Cost to deliver + Cost to return
Supply chain asset management	Cash-to-cash cycle time	Inventory days of supply + Days of receivables outstanding – Days of payables outstanding

Reliability

In the category reliability as applied to the delivery process for example, you can look for waste in terms of shipping the correct product to the correct place and customer at the correct time. This includes looking at whether you have shipped the product in perfect condition and packaging, in the correct quantity, and with the correct documentation. The resultant metrics measured would include the following:

- **Delivery performance**—Did it both ship and deliver to the client when the client originally wanted it? Some companies adjust the delivery date based on availability, change the date in their system, and measure performance based on the new delivery/promised date. This results in an inaccurate view of delivery performance.

- **Order fill rate**—It is important to know whether an entire customer order shipped complete. This metric is typically a lower percentage performance than line item fill rate, which should also be measured.

- **Accurate order fulfillment (at various levels of detail)**—This quality measurement looks at shipping errors such as the wrong order or item(s) shipped to the customer (or the wrong quantity of requested items).

Responsiveness

Responsiveness measurements relate to how quickly your supply chain and logistics function can deliver products to the customer. They can include measurements such as order fulfillment lead time, transit times, on-time delivery, and even overall cycle or dock-to-dock time (i.e., the total time key material sits in a facility, which is a good measure of your organization's leanness).

Flexibility

Flexibility is a measure of your supply chain's agility and response time when there are changes in the supply chain. As you know, many unanticipated changes due to economic, environmental, political, and other issues can occur, and being flexible enough to deal with them can give you a competitive edge.

Cost

It is, of course, important to manage your supply chain and logistics costs as they are signs of potential areas of waste. These measures include cost of goods sold (COGS), total supply chain and logistics cost (in dollars and as a percentage of revenue), transportation and distribution costs, warranty/returns, and a host of other individual costs.

Asset Management

Asset management metrics look at how effectively a company manages assets—including fixed assets and working capital—to meet demand. Metrics include order-to-cash cycle, inventory, and asset turns.

Supply Chain Analytics

A very popular topic today is data analytics, which is the science of examining raw data to help draw conclusions about information. When applied to the supply chain, it is often called *supply chain analytics*. (See Figure 18.1 for a high-level view of supply chain reporting and analytics.) In many industries, companies and organizations use analytics to drive insight, make better business decisions and actions, and verify (or disprove) existing models or theories.

Figure 18.1 Reporting and Analytics in the Supply Chain

While data analytics is not an entirely new concept, advancements in technology have made the gathering, filtering, and analysis of data much easier, more efficient, and more affordable.

One way to look at data analytics is to break it into four categories:

- **Descriptive analytics**—Also described as business intelligence (BI) systems, descriptive analytics uses historical data to describe a business. In supply chain, descriptive analytics helps better understand historical demand patterns, understand how product flows through the supply chain, and understand when a shipment might be late.

- **Diagnostic analytics**—Once problems occur in the supply chain, an analysis needs to be made of the source of the problem. Often this can involve analysis of the data in the systems to see why the company was missing certain components or what went wrong that caused the problem.

- **Predictive analytics**—It is often possible to use data and statistics to predict trends and patterns. In the supply chain, predictive analytics can be used to forecast future demand or to forecast the price of a product.

- **Prescriptive analytics**—This type of analytics uses data to select an optimal solution. In the supply chain, you might use prescriptive analytics to determine the optimal number and location of distribution centers, set inventory levels, or schedule production.

Traditional, more static measures tend to be based on historical data and not focused on the future. They don't relate to strategic, nonfinancial performance goals such as customer service and product quality and don't directly tie to operational effectiveness and efficiency.

Every businesses that has a supply chain devotes a fair amount of time to making sure it adds value, but new advanced analytic tools and disciplines make it possible to dig deeper into supply chain data in search of savings and efficiencies.

The supply chain is a great place to use analytic tools to look for a competitive advantage because of its complexity and also because of the prominent role the supply chain plays in a company's cost structure and profitability. Supply chains can appear simple compared to other parts of a business, even though they are not. If you keep an open mind, you can always do better by digging deeper into data as well as by thinking about a predictive instead of reactive view of the data.

Supply Chain Decision Support and Analytics Technology

Traditional supply chain management systems, while great for automating operations, don't necessarily feed critical decision-making loops that become more numerous and frequent as a manufacturing company evolves into a more sophisticated and competitive business. This is a result of SCM/ERP systems reflecting primarily what has already happened instead of what is or will happen, and it prompts the need more sophisticated analytics and decision-making technology.

Today's supply chain analytics best-practice software may include the following types of functionality:

- Monitoring, controlling, alerting, simulating, and measuring critical supply chain events

- Moving your organization from a reactive mode to a proactive position

- Leveraging human assets to do more with less
- Managing supply chain processes on an exception basis
- Focusing on the highest-priority tasks and activities
- Resolving supply chain issues before they become problems

Adding analytical capability can provide functionality that yields better, more informed decisions.

In an *IndustryWeek*/SAS survey, only about 12% of the manufacturers responding were satisfied with their SCM and ERP systems' capability to analyze relevant data for timely decision making and reporting. The survey showed that clearer visibility into operations, marketing, and consumer activity through supply chain analytics can better predict challenges and respond to them proactively, thereby increasing both efficiency and profitability [W/SAS, 2010].

On the pure technology side, the survey found the following:

- **Efficiency and performance gains require predictive, data-driven insights—**Supply chain analytics technology can help improve forecasting accuracy, understand demand patterns, optimize supplier performance, and reduce finished goods inventory and stock-outs.

- **Traditional SCM/ERP systems are not advanced enough for current economic conditions—**Most successful companies realize that they cannot simply rely on surface-level data from various transactional systems. Integrating data from transactional systems with downstream consumption data as well as upstream supply data removes inaccuracies and provides forward-looking analytical insights. Using supply chain analytical tools can help you discover trends, anticipate events, and understand the underlying drivers of costs and revenue, allowing your organization to be innovative and agile in a rapidly changing business environment.

- **Analytics are the future for next-generation supply chains—**Supply chain executives expect their future systems to help them make more strategic decisions, including how to control costs, improve demand forecasts, and upgrade customer service. Traditional supply chain systems have not successfully addressed these issues due to their limited ability to answer only very basic questions. Next-generation supply chains will require advanced analytical capabilities featuring constraint-based optimization, advanced forecasting, what-if analyses, scenario planning, business simulation, and modeling. [W/SAS, 2010]

Currently, "big data" and advanced analytics tools are being integrated with network optimization, demand forecasting, integrated business planning, supplier collaboration, and risk analysis software tools. Some of the top supply chain analytics software solution providers include SAP, Oracle, JDA, Manhattan Associates, SAS, IBM, and Logility, and these solutions are typically offered as modules of SCM and ERP systems and in some cases as standalone analytical systems, either on an installed basis or as cloud-based on-demand systems.

Lean and Supply Chain Analytics Technology Case Studies

The following sections provide some actual examples of companies that have used supply chain analytics tools and technology to improve to take their supply chains to the next level.

Case 1: Business Intelligence in the Supply Chain—Better Decisions Through Data for Anna's Linens

Challenge

Anna's Linens, based in Costa Mesa, California, is a family-run business with more than 3,200 employees that operates around 320 stores in more than 20 states. The company wanted to optimize performance across its supply chain.

Approach

Anna's Linens selected a business intelligence portal from Transplace to access a suite of reports and dashboards that provide clear visibility to transportation performance. This portal would give the company access to more than 50 different transportation metrics and KPIs, with graphs showing trending visibility and radial dials displaying performance based on type of data, data elements, and how they are analyzed.

The business intelligence data would need to be delivered in the form of real-time dashboards and reports, making it easier for the supply chain team to use the data to make operational decisions more quickly and efficiently. The depth and frequency of data available from Transplace would provide the necessary basis for process decisions and strategy evaluation.

Results

Anna's Linens is now able to monitor performance to make sure it has improved and to see if it has incurred any costs associated with the change. The system has also helped the company avoid overreacting when issues arise that don't necessarily signal a larger problem in the supply chain. As a result, Anna's Linens now has actionable visibility to performance and cost information, which has enabled the company to optimize the results while at the same time devoting less time and resources to sifting through a lot of data [Partridge, 2013].

Case 2: Steel Manufacturer Improves Performance, Profitability

Challenge

A large Asian steel manufacturer with 19,000 employees wanted to update its 30-year-old business practices to improve efficiency and competitiveness.

Approach

The company decided to base two of its process innovation (PI) programs on SAS software. First, it used SAS software to extract, transfer, and transform its ERP and legacy data into a data warehouse, allowing data to be compared on an "apples to apples" basis and checked for quality. Next, the company combined SAS's analysis capabilities with its Six Sigma project tracking system in order to gather data on PI projects, identify any critical quality issues and analyze them for root causes, resolve issues early, and improve overall manufacturing processes.

Results

The steel manufacturer had a 50% reduction in lead times for standard hot coil production (from 30 days to 14 days) and a 60% reduction in inventory (from 1 million tons to 400,000 tons). In addition, by analyzing and making improvements to its manufacturing process, the company was able to reduce the scrap ratio on hot coil from 15% to 1.5%, for total savings of over $15.5 million in less than two years [W/SAS, 2010].

Case 3: H.D. Smith: Why Analytics Is Eating the Supply Chain—New Tools Are Helping Companies Achieve the Supply Chain's "Holy Grails"

Challenge

Pharmaceutical wholesaler H.D. Smith found that data to predict and meet demand was dispersed across multiple subsidiaries and storage systems. For management, getting everything in one place for analysis was a time-consuming challenge. The company was spending a lot of time trying to pull data out to determine the current demand on volatile items, and by the time they could access it, things had changed. At the time, H.D. Smith was using Microsoft's Access and Excel tools for analysis.

Approach

H.D. Smith selected an analytics platform from FusionOps to give new visibility to the supply chain, to better anticipate and meet demand, and to offer service levels it couldn't have previously offered.

The company began rolling out FusionOps after running pilot tests in the fall of 2013. The software has modules for procurement, finance, inventory, sales, production planning, and customer service. It also has a relatively recently added prescriptive analytics functionality, which enables users to simulate different scenarios for a better understanding of the pros and cons of alternative decisions. The analytics functionality allows management to create dashboards every morning to guide their efforts for the day.

Results

By segmenting inventory according to demand and other characteristics and focusing on high-value, high-volume, and high-urgency items, H.D. Smith can now keep inventory better aligned with what the company will actually sell and when.

It has achieved an actual reduction in inventory on hand and an improvement in service levels. In fact, it now beats its 98.5% service level target every month, whereas in the past it would hit that target only two months out of the year [Noyes, 2016].

As we know, the supply chain is actually more like a web than a chain; it is a complex network of customers, suppliers, and partners. Managing and having visibility into this global network that is continually changing requires extensive collaboration, the topic of Chapter 19, "Collaborative Supply Chain Systems."

Part VIII

Where Do We Go from Here?

19

Collaborative Supply Chain Systems

By integrating and collaborating with partners and customers downstream as part of a Lean strategy for your supply chain, you actually open a window into your future—and even your past. It's like having the ability to time travel. Before we travel through time, however, let's look at the subtle differences between integration and collaboration.

Integration of supply chain components started in the 1970s, when electronic data interchange (EDI) created a business-to-business communications standard, followed in the 1990s by enterprise resource planning (ERP) systems with common databases. Then came the introduction and growth of the Internet.

However, true supply chain collaboration is more than just integrating information among business functions and partners. It involves companies working together to improve data sharing and is an interactive process that results in joint decisions and activities—often in multi-company teams made up of individuals from various disciplines in each organization.

Furthermore, supply chain collaboration is not easy to accomplish for many reasons, including a tendency to rely too much on one technology, failure to understand when and with whom to collaborate, and a propensity for distrust among partners.

In fact, many collaboration attempts through the years have failed; as many as 8 in 10 can fail. Reasons range from lack of commitment from senior management, to failure to provide sufficient resources to make collaboration efforts work, to limited resources spread too thinly over many initiatives. Another problem may be that initiatives are spread over two different organizations. However, collaboration is well worth the effort as it can result in reductions in inventories and costs, along with improvements in speed, service levels, and customer satisfaction.

Now that we're clearer about supply chain integration versus collaboration, let's talk time travel.

Collaborative programs, have been around since the late 1980s. They include quick response, efficient consumer response, vendor-managed inventory (VMI), and now collaborative planning, forecasting, and replenishment (described later in this chapter). These programs all involve getting a more accurate downstream picture of the supply chain, using information such as point-of-sale data, retail store and distribution center inventory balances and withdrawals, and current and future events such as promotions, discounts, or advertising. These types of solutions reduce the bullwhip effect (i.e., progressively larger inventory swings in response to changes in customer demand) and therefore also reduce supply chain volatility, inefficiency, and waste.

Through a structured integration and collaboration process, it is possible for manufacturers and distributors, in essence, to time travel and see potential causes of future disruptions before they occur. While an initial investment in resources may be required, opportunities for fewer stock-outs on store shelves, upselling, and cross-selling may make the investment worthwhile.

The 80/20 Rule

The initial investment often causes companies to shy away from integration and collaboration programs without looking at the big picture. One way to justify the investment is via the Pareto principle, also known as the 80/20 rule. The rule states that in business, there is a natural tendency for a small number of items to generate a disproportionately large portion of sales and/or profits.

The 80/20 rule also applies to an organization's customer base: An organization should focus on integration and collaboration efforts with the large customers that make up the greatest portion of sales. Companies can use the advance information gained through this process to significantly improve forecasts, thus boosting service levels, reducing inventory costs, and removing other types of additional waste in the supply chain [Myerson, 2014].

Throughout most of this book, we have concentrated on technology and systems that help with internal integration and collaboration. In this chapter, we look more at external integration and collaboration with customers and suppliers/partners as a way to truly enable a Lean global supply chain.

Collaboration for a Lean Supply Chain

In today's global economy, with shorter lead times and product life cycles, as well as volatile demand, it is especially critical that we have timely visibility both downstream and upstream in our supply chain in order to be flexible and agile.

In general, collaboration enables you and your supply chain partners to do the following:

- Improve forecast accuracy by getting closer to the points of demand and supply
- Strengthen strategic supply chain relationships and profitability
- Enhance sales and operations planning to achieve corporate goals
- Accelerate and manage demand plans, direct material procurement, and fulfillment throughout the supply chain
- Manage supply chain processes on an exception basis
- Resolve critical supply chain events through automated monitoring and alerts

Collaboration with customers and suppliers accelerates sales and operations planning (S&OP) as well as strategic trading partner relationships to manage demand plans, direct material procurement, fulfillment, and financial goals to increase profitability and improve service.

Customer Collaboration

Customer collaboration involves receiving demand signals and automatically replenishing the customer's inventory based on actual demand. This is seen primarily in consumer products and other industries that have downstream distribution systems that extend to retailers.

This type of integration and collaborative effort enables manufacturers to shift from a "push" system to a demand "pull" supply chain, while combining both forecasts and actual customer demand.

Collaborative continuous replenishment processes such as *quick response* (*QR*) and *efficient consumer response* (*ECR*) are more responsive than purely forecast-based processes. They are driven largely by actual customer demand and also provide visibility in out-of-stock situations so that manufacturers and retailers can react more quickly. Point-of-sale (POS) information can add visibility across the entire supply chain as well when included in a collaborative replenishment process.

Another type of customer collaboration that focuses on forecasts is known as Collaborative Planning, Forecasting, and Replenishment (CPFR, which is a trademark of the Voluntary Inter-industry Commerce Standard Association [VICS]). It is an outgrowth from some of the earlier customer replenishment initiatives such as QR and ECR. In general, CPFR is an attempt to reduce supply chain costs by promoting greater integration, visibility, and cooperation between trading partners' supply chains. It combines the intelligence of multiple trading partners in the planning and fulfillment of customer demand.

Figure 19.1 shows collaborative or vendor-managed inventory configurations in terms of the level of sophistication or complexity. Levels 1 and 2 have been implemented in various industries and would programs such as QR and ECR. Levels 3 and 4 are more advanced and include CPFR-like programs.

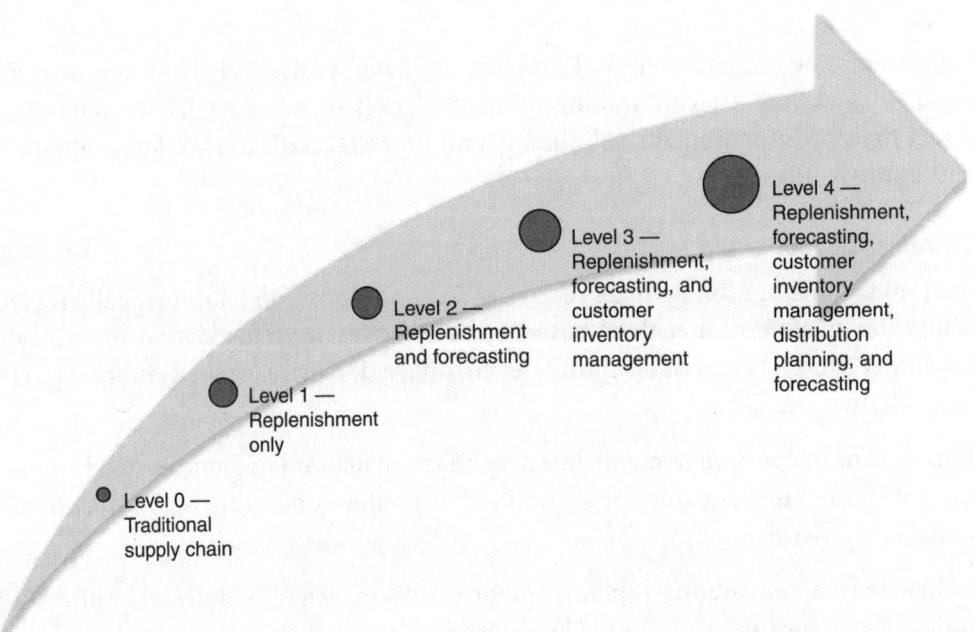

Figure 19.1 Types of Collaborative or Vendor-Managed Inventory Configurations in Supply Chains

Supplier Collaboration

The following are some of the types of supplier collaboration:

- **Kanban**—This signal-based replenishment process used in Lean or just-in-time (JIT) production uses cards or other visual signals such as a line on a wall to signal the need for replenishment of an item. Using collaborative technologies, the kanban process allows customers to electronically issue the kanban replenishment signals to their suppliers, which can then determine requirements and see exceptions.

- **Dynamic replenishment**—In this process, suppliers compare customer forecasts or production schedules with their own production plans to match supply and demand. It allows suppliers to adjust to changes in customer requirements or supply shortages.

- **Invoicing processes**—Automating invoicing and related processes gives the visibility to the vendor for the entire supply side, including purchase orders, releases, supplier managed inventory, kanbans, and dynamic replenishment.

- **Outsourced manufacturer collaboration**—When managing outsourced manufacturing relationships or contract manufacturers, you must shift your focus from owning and organizing assets to working collaboratively with partners.

  The collaborative efforts should help simplify processes such as product development and should also reduce manufacturing costs and improve reaction to response to customer demand. Any efforts to automate these processes should support information sharing, collaboration, and monitoring activities that are needed to effectively manage the relationship with a contract manufacturer.

- **Logistics service partners**—Collaboration can increase the effectiveness of any partnership, but it can be essential in maintaining an effective distribution network. Outsourced transportation, warehouse, and third-party logistics companies have become essential parts of many firms' supply chains. Collaboration involves sharing information from an assortment of transactional functions, including contracting, tracking, billing, auditing freight bills, and so on and, more strategically, sharing and collaborating to support the planning and execution side for improved customer service, flexibility, agility, and cost reduction. Technology has increased the ease with which companies can monitor the supply chain. However, increased focus on social platforms can improve the communication between businesses and their logistics partners, making distribution networks faster and more responsive. We will at this in more detail in Chapter 20, "Emerging Technologies and Their Potential Impacts on the Lean Supply Chain."

Integrated and Collaborative Technology for a Leaner Supply Chain

Many of the systems discussed throughout this book, such as ERP, forecasting, and DRP systems, while traditionally viewed as internal integration and collaboration tools, in many cases can also help with external integration and collaboration.

An example of this is the use of a combination of EDI/Internet, forecasting, and DRP software in an ECR program for a manufacturer that also imports customer data such as POS at the retail level and on-hand inventory, shipments, and open purchase orders at the retailers' distribution center locations. All this helps ensure that the company's products are always available at retail locations.

Some technologies, such as EDI, can be used with an unlimited number of partners as they are more transaction-based and fairly automated. On the other hand, others, such as forecasting and inventory planning, require a more strategic and selective approach because they require more commitment of resources. (Remember the 80/20 rule mentioned earlier in this chapter.)

Collaborative software helps bring previously disconnected entities onto the same page, especially with the increasing use of cloud computing. Vendors are developing solutions that help optimize logistics and procurement, enable the exchange of data, and encourage collaboration among business partners. This has enabled many global manufacturers, distributors, suppliers, and retailers to be on the same page, in real time, and at any given time.

In fact, research by the University of Tennessee's Global Supply Chain Institute found that the following:

> Organizations are investing in B2Bi (business-to-business electronic integration) to cut costs and increase business flow efficiency. In the survey, 94% saw great improvement in their electronic connectivity capabilities and 68% said their clients found it easier to do business with them after using cloud-based B2Bi managed services.
>
> The study also found that there are more opportunities for streamlining processes than the companies have resources to do it. This is a result of as increased

pressures on internal IT departments to meet basic business objectives. 96% of those surveyed said they are linked electronically with at least one of their trading partners, yet the average organization spends only 5 percent of its IT budget on electronic connections. Electronic connections are expected to increase more than 20% over the next three years with 69% of the company's surveyed saying that they intend to increase the number of customers they trade with electronically.

The study presents several examples of successful collaborations. For example, an office supplies retailer invested time and technology to collaborate more effectively with a major supplier. Afterwards, their in-stock fill rates rose significantly to nearly 99%, lead times were reduced nearly 60%, forecast accuracy was improved 30% and inventory turnover increased 9%. [www.sdcexec.com, 2016]

Lean and Collaboration Technology Case Studies

Collaboration relies more on systems integration than does the functional software discussed previously in this book. The following are some cases that show how finding creative ways to connect partners in a global supply chain can specifically enhance a business's productivity and bottom line.

Case 1: Agilent Technologies—Comprehensive Supply Chain Visibility Across a Multi-Enterprise Supply Chain

Challenge

Agilent offers a broad range of innovative measurement solutions in the electronic test, life science, and chemical analysis markets. It has a large mix of products in a low-volume environment, which creates a lot of variability in its plans. Products are typically configures to order, along with many other options.

Outsourcing of manufacturing operations is a key strategy for Agilent. This means the company has to make its way through demand and market volatility, while at the same time coordinating a virtual supply chain network made up of multiple supplier and contract manufacturing (CM) partners. Therefore, Agilent requires visibility from the CM partners on everything from demand and supply, to commit plans, to delivery status.

Overall, Agilent requires accurate, synchronized, and timely information and a common view of that information for itself and for all of its supply chain customers, suppliers, and partners.

Challenge

Agilent's existing requirement-generation process took several weeks, depending on how complex the relationship. When there was a change, it had to flow through the same process, and that also took several weeks to reach the lowest level of the supply chain. By the time that level had responded, things had changed again.

Agilent's challenge was to flow information down fast enough so that everyone could respond to the appropriate signal quickly. To meet this challenge, in late 2010, Agilent began a supply chain improvement program that involved creating a vertically integrated planning process to consolidate all of its different MRP systems and those of its CM partner, to create a single plan (via bill of material integration) with minimal data lag.

Agilent selected RapidResponse from Kinaxis their platform to merge all its information systems and to provide the supporting capabilities to create and manage a single vertical supply chain.

Results

RapidResponse created a vertically integrated plan extending from top-level demand and exploded through the BOM and through all the sourcing. This plan allows Agilent to see demand going through to the CM partners (and back through Agilent for intersourcing arrangements).

Agilent now has one common view where it can trace and facilitate information flow from the top to the bottom of the multi-enterprise supply chain—all from within one system. It now has the following:

- Integrated information with minimal delay, which has improved information flow and decision-making times
- Complete information, fast and deep simulation capabilities, and a better understanding of all the interdependencies enabling better decision making
- More timely and accurate answers to provide to customers
- Increased business opportunities and minimized risks
- Alignment of actions with corporate targets

For example, Agilent was able to reduce its new order assessment and commitment process to 1 or 2 days, down from 3 to 14 days with manually reconciled information. It can now immediately assess an order and get back to the customer with a commit date and production plan within 1 or 2 days [Kinaxis Customer Spotlight, 2014].

Case 2: Arrow Electronics Automates Replenishment Program with One Network's Real-Time Value Network

Challenge

Arrow Electronics is a $23 billion company that specializes in distribution and value-added services related to electronic components and computer products. Over the years, Arrow has utilized its Customer-Automated Replenishment System (CARES) to monitor the inventory of components at customer sites. CARES was running on computers in each customer's manufacturing plant. Furthermore, the legacy CARES system didn't provide all the functions Arrow required.

Arrow's customers used CARES in connection with either a kanban or min/max replenishment policy, in which inventory falling below a minimum level triggered replenishment to the maximum level. It also no longer made sense to ask customers to load the software on their own computers, which required additional support by Arrow.

The company had considered building a new system in-house since it wanted to convert customers to a system they could access over the Internet so that Arrow could become more demand driven. It ultimately decided that it could reach its goals faster and less expensively by using an external vendor.

Approach

Arrow selected One Network Enterprise's real-time value network. One Network has thousands of business partners, and it allows clients to build their own modules, use their existing legacy systems, or extend One's network to integrate to almost anything (e.g., SAP, Oracle) and unlock their legacy systems for collaboration with other organizations.

Most importantly, Arrow chose One Network because of that company's focus on building networks to manage replenishment and other supply chain functions. One Network had created several electronic communities that successfully automated transactions and allowed collaboration among trading partners.

Results

Agilent went live with One Network in three months. It now gives all of its supply chain partners access to important information, such as inventory levels and actual demand at any time via the Internet. Orders are automatically fed into Arrow's system upon receipt.

The system supports new functionality such as the ability to accommodate multiple manufacturing locations and multiple manufacturing cells within one location.

Arrow has also gained more visibility into its supply chain so that it can anticipate problems. In the past, they had to wait until a customer needed a part, and then it often found that the part wasn't available.

In the future, Arrow will be able to use this network to see when customers sell end products that contain Arrow's components and gain visibility into how those components move and are stocked through manufacturing and logistics processes.

Arrow's network now accommodates hundreds of companies and about 60 original equipment manufacturers (OEMs) and suppliers that serve those OEMs. The more suppliers on One Network, the more value the system has to the OEMs [One Network Enterprises, 2011].

While the various technologies we have described throughout this book can be used to improve the efficiency and agility of a business's global supply chain network, it is evolving rapidly. In Chapter 20, we discuss where and how emerging technologies such as mobile solutions, cloud computing, and speech recognition can work hand in hand with ever-changing processes driven in part by the continued growth of the Internet, omni-channel marketing (including e-commerce), and social media to achieve even greater process improvements.

20

Emerging Technologies and Their Potential Impacts on the Lean Supply Chain

T he tools and methods to communicate, gather, analyze, and act on information have multiplied in recent years, but with that explosion have come greater expectations from customers who want products and services customized to their needs and delivered quickly and as promised.

Much of the technology used to plan and deliver products and services today is still evolving and has not yet reached its full potential. Take, for example, the recent development of embedding chips in credit cards (known as "chip," or "EMV," cards) to potentially add speed, accuracy, information, and security to the retail checkout process. However, while there was quite a buzz about those cards in late 2015, many retailers delayed their implementation until after the 2015 holiday season because the cards can take 10 to 20 seconds to process, and they were afraid that would create a bottleneck in the checkout process.

As another example, Walmart intended to roll out the required use of RFID to all suppliers in 2003. It eventually scaled back to its top 300 suppliers. While this transition has taken longer than anticipated, Walmart's involvement has helped the technology to develop further. Walmart, Macy's, and other retailers have now started using it in stores to manage inventory accuracy and replenishment from the back room and distribution center.

While some technology fails and other technology take longer to succeed, there is no doubt of its impact now and in the future on the supply chain.

General Supply Chain Trends

It is helpful to look at what is going on in general with the supply chain. This can then help determine which of the many technologies (hardware and software) can be most helpful to your company.

The following are some relevant predicted trends:

- **Advantage to networked companies**—Collaboration is a critical element of a supply chain, and more companies are fully integrated throughout their network of suppliers and customers, with everyone having access to the same latest information. It is best when those who are collaborating have a shared mission and strategies to get ahead of the competition. This allows a business network to be agile, responsive, flexible, and efficient … all Lean traits.

- **Avoiding a "silo" mentality**—It is critical to eliminate silos by focusing on transforming efficiency, transparency, and exposure to risk to reduce delays and react more quickly to changing trends.

- **Responsibility and traceability**—When operating globally, there is a need for transparent and traceable supply chains, enabled by safe and secure IT platforms.

- **Sustainability**—Consumers are becoming increasingly aware of sustainability-related issues, so companies must have policies in this area, for both forward and reverse logistics processes.

- **Acceleration of the supply chain**—This can be best achieved through partnerships and collaboration. Technology and supply chain partners are starting to offer strategic, tailored, and easily deployed packages to help businesses accelerate their supply chains.

- **Digitization of the supply chain**—Supply chain analytics systems that are now readily available allow companies to more actively develop and manage key performance indicators (KPIs) and create innovative ideas through monitoring, sharing, and interpretation of data.

- **Big data**—Businesses that gather and analyze big data from shipment and transportation continue to improve their efficiency, especially when they use collaborative solutions. The results can be improvements in cost reduction, capacity control, and risk management.

- **Disruption risk**—Weather, terrorism, political, financial, energy, and many other types of risk continue to affect the global supply chain. It is critical to proactively assess the risk of disruption and develop a risk management strategy to be successful in the long term.

- **Supply chain at the C-level**—Executives, investors, and board members now better understand the value of a good supply chain and its inherent risks. The supply chain will continue to be prominent as companies try to balance growth, profitability, responsibility, and customer service.

- **Social media**—Social media has changed the way we communicate and collaborate with our personnel and how businesses network—beyond ERP and portals to work with trading partners. It uses a "one-to-many" approach, where information is delivered in real time to thousands of contacts in a network [Felgendreher, 2014].

Supply Chain Software and Connectivity Technology Trends

As supply chains have become more globalized, software has become even more important to a company's success. More and more companies—especially small- to medium-size enterprises (SMEs), many of which rely on manual SCM processes—are now realizing that investing in SCM software both produces a significant return on investment and enables and improves business processes.

Rapid advancements in supply chain software solutions, especially in the areas of external integration, centralized data and control, mobile computing, and the cloud, are enabling companies, small and large, to become more connected, efficient, and flexible.

The following are some significant supply chain software trends:

- Enterprise resource planning (ERP) continues to expand in scope, adding increased supply chain visibility both upstream with suppliers and downstream with customers. ERP systems are integrating with manufacturing execution systems (MES), job scheduling processes, and job status tracking within the plant, and some are even communicating with machines, using an increasing number of sensors and data feeds, for more accurate time and costing, as well as for predictive scheduling.

- Command and control centers accumulate supply chain information centrally for collaboration purposes for fast response to problems locally and globally.

- Mobile devices (i.e., phones, tablets, and laptops) are reducing the need for paper-based information as truck drivers with tablets connected to ERP can get digital signatures and time stamps when making deliveries. Sensors and automated feeds track goods on the road down to the item level and instantaneously report on equipment breakdowns and defects, as well as provide other notifications. Databases and analytic engines can be accessed from anywhere.

- Cloud-based hardware and software have enabled much of the information sharing that is accelerating supply chain flow, making comprehensive supply chain solutions affordable for even the smallest manufacturers, distributors, and retailers. Software-as-a-Service (SaaS) platforms allow small manufacturers to "pay as they go" to use full-featured software without the significant investment of time and money needed to develop and maintain homegrown systems. Even in the case of larger companies, SaaS solutions eliminate the overhead cost of system maintenance and upgrades.

Furthermore, investing in training and development of the people who select and maintain the software, analyze data, or use data to manage the supply chain is critical in getting and maintaining a competitive advantage [PNC Bank, 2016].

Supply Chain Software Best Practices

Supply chain software vendors are building and integrating many of the following best practices into their solutions, and these practices also lend themselves to a Lean supply chain:

- **Demand-driven supply chains**—Demand is a key driver for the use of working capital, operational capacity, and resources, and of course supply chain planning. Many of the best-performing organizations segment their customers and focus on the value proposition for each segment, while managing demand to meet the goals of each segment. SCM software allows planners to create and manage forecasting and fulfillment strategies by product grouping as well as to simulate buying patterns and perform what-if analysis to align the supply chain with the market and the organization's strategies and goals.

- **Dynamically synchronized supply chains**—Managing multiple supply chain strategies based on demand types is a fairly common best practice these days. Software can support simulation, analysis, and planning in the selection of supply chain strategies aligned with different products or product groupings. Software can also monitor for changes in a product's demand type while dynamically allowing for changes in SCM strategies, such as those that focus on cost, time, quality, and flexibility (discussed in Chapter 1, "Lean Supply Chain and Technology: A Perfect Combination"), all the while supported by an underlying Lean philosophy.

- **Globalization**—To minimize the challenges and enhance the benefits of globalization, new SCM systems need real-time updates using GPS, RFID, and other types of integrated tracking and identification systems, interconnected supply chain partners providing real-time visibility to product movement, and collaboration on planning, execution strategies, and mitigating risks. Best-practice supply chain systems today provide some degree of these important capabilities, but they will continue to evolve and improve.

- **Risk management**—To manage risk, processes need to be standardized with the appropriate amount and type of data and technologies. Many supply chain software vendors have added functionality to better understand, manage, and mitigate risks. Systems can support this through the use of simulation models, real-time updates on events, and facilitation of contingency plan implementation. For example, if a container is lost during shipping, a notification can show in the transportation system so the forecasting team can shift demand and reprioritize or reallocate fulfillment.

- **Collaborative relationships**—As discussed throughout this book, supply chain systems are providing new ways to facilitate relationships and collaboration with customers, suppliers, and other partners externally and internally in a way that crosses hierarchical, departmental, and organizational silos. Examples include S&OP software that coordinates and includes input from internal and external members of the organization's supply chain and enhanced collaboration through integration of supply chain management systems with customer relationship management and supplier relationship management applications.

Supply chain management is becoming increasingly strategic, adaptive, and demand driven as opposed to primarily focusing on operational efficiencies. As a result, software vendors are supporting this strategy by giving users more flexibility in deploying new best practices and taking advantage of additional trends such as cloud computing, mobility, and social media [Heistand, 2016].

There is even a trend toward the "Uber-ization" of the trucking industry. As you undoubtedly know, Uber is a technology platform that connects driver-partners with riders through a smart phone app. The somewhat logical next step might be to "Uber-ize" the motor carrier industry, as it fits the profile very well since the majority of U.S. freight moves via truck, yet the market is very fragmented. On top of that, many companies (especially small- to medium-size companies) use transportation brokers, which charge up to a 20% commission to line up a carrier. There has been a rush to develop Uber-like apps to cut out the brokers or, in some cases, make their jobs easier. The apps could lower expenses, improve visibility across a company's entire supply chain, and, most importantly, prevent trucks from having to move while empty. A few examples of these apps that are currently available or in development include the following:

- **Transmission**—This platform connects freight requestors (shippers, brokers, freight forwarders) with their preferred carriers to reduce time spent covering, tracking, and delivering freight. It concentrates on creating a private network of partners.
- **Cargomatic**—This app being developed for local shipping targets shippers and local trucking companies with full or partially empty vehicles.
- **Keychain Logistics**—This technology helps commercial truck drivers find freight from shippers and brokers. It enables shippers to directly book freight carriers, bypassing broker commission with a network of tens of thousands of carriers all over the United States.

While some of these apps may end up being more successful than others and others will surely come along in the future, there is certainly room for a product that automates processes and fills empty trucks with a more user-friendly pricing policy [Abrosimova, 2014].

Supply Chain Hardware Technology Trends

While there tends to be a lot of focus on software, a platform for the software and ways to gather, store, process, and deliver information require hardware. Therefore, we now look at some trends and technologies that are impacting supply chain operations in production, distribution, retail, and remote services:

- **Comprehensive connectivity**—Various forms of wireless connectivity are available today, including Bluetooth, wireless local area networking, and cellular wide area wireless networks for voice and data communication.

  The greatest impact is achieved by combining these technologies into single devices that provide multiple forms of wireless functionality. While smart phones are ubiquitous and provide convenient voice and data access, they are limited for delivery, field service, and other mobile supply chain operations because the screens and interfaces aren't optimized for enterprise applications, and they are not rugged enough for industrial usage. More common in industry has been the use of ruggedized handheld computers for more reliability, but these devices lack cell phone capability.

- **Advanced wireless (voice and GPS)**—There are now certified rugged cellular handheld computers for voice communication that combine data collection, data communication, and cell phone functionality on one device.

Converging data and voice onto one device can significantly reduce administrative support. For example, Stanley Steemer, a carpet cleaning franchise, automated route operations by using mobile computers with integrated wide area wireless connectivity, GPS, and a magnetic strip reader to process credit card payments in real time when service was finished. This change allowed the company to eliminate a full-time dispatcher position at each branch and has greatly reduced the time needed to complete and process paperwork.

- **Speech recognition**—Speech recognition enables productivity and has typically been implemented for high-throughput operations such as those in warehouses. While bar code data entry is somewhat more accurate than speech recognition software, speech recognition software tends to be much more productive.

- **Digital imaging**—Transportation and distribution companies use digital cameras in mobile computers to capture proof of delivery, stamped invoices, and other delivery information.

For example, Mission Foods went from manual invoicing to creating and printing invoices on a handheld computer with a mobile printer for the customer at delivery. The invoice records are sent to headquarters in real time via the cellular network. Mission Foods no longer has to scan thousands of paper invoices. This demonstrates the convergence of wireless connectivity.

- **Portable printing**—Mobile printers and computers help sales, service, and delivery people give customers documentation "on the spot" while creating an electronic record to eliminate the need to process paperwork. Mobile printing also delivers labor savings in warehouses, for example, by allowing workers to avoid excess travel to pick up printed labels, pick tickets, manifests, and so on.

- **2D and other bar coding advances**—Auto-focus imaging technology will help bring 2D bar codes (better known as quick response [QR] codes) into the mainstream for inventory management, traceability, MRO inventory, and other operations. QR codes are two-dimensional (2D) barcodes that look like squares or rectangles and contain many small, individual dots. A single 2D barcode can hold a significant amount of information, and many are readable when printed at a small size or etched onto a product. There are also new bar code readers that can read linear and 2D bar codes from inches up to 50 feet away.

- **RFID**—RFID has become more practical with the use of asset management and supply chain operations. Vehicle-mounted and other mobile RFID readers eliminate the need to purchase, install, and maintain separate RFID readers at each dock door.

    The U.S. Social Security Administration (SSA) implemented RFID systems in a warehouse to track inventory and improve the efficiency of shipments to branch offices. As a result, it saw a 39% productivity improvement and $1 million in annual savings.

    While compliance tagging and other RFID supply chain initiatives from companies such as Walmart get most of the attention, research shows that implementation is growing even faster for asset management.

- **Real-time location systems (RTLSs)**—RTLSs allow for the expansion of wireless local area networks to be used as asset tracking systems. One example of this type of usage is to track forklifts via their vehicle-mounted computer's radio [www.dynamic-systemsinc.com, 2007].

- **Advanced robotics**—A new generation of advanced robots with enhanced sensing capabilities and algorithms can better sense their environment and make decisions based on changes in that environment.

    For example, products in a warehouse are different sizes and shapes and may not be positioned the same way or in the same location each time they are handled. New vision-sensing technologies are helping robots adjust to these variations to perform mixed-case palletizing and de-palletizing, for example. Robots in material handling can eliminate the need for humans to perform routine tasks and add greater speed and accuracy to repetitive tasks, helping reduce material handling costs.

- **Autonomous vehicles**—These vehicles will be capable of making decisions in response to their environment and will offer far more utility and flexibility than the automated guided vehicles (AGVs) used in warehouses today, which follow preplanned routes and can't navigate around obstacles. Autonomous cars are currently being developed, and autonomous forklifts will most likely follow at some point.

- **Alternative energy**—Lighting represents a large percentage of the total electricity cost for distribution centers. Some large distribution centers are seeking out alternative energy sources to reduce these costs. For example, the Toys R Us distribution center in Mt. Olive, New Jersey, has equipped its roofs with solar panels and offset 72% of the 1.3 million square foot facility's electrical needs.

  Within a warehouse, the energy efficiency of forklifts and automated storage and retrieval systems has kept improving, in large part due to new forms of power regeneration and new approaches to monitoring and balancing performance and energy usage. More environmentally friendly and efficient hydrogen fuel cells and lithium-ion batteries are being applied to forklifts and have shown some promise compared to traditional lead-acid batteries. The lithium-ion batteries are more efficient than lead-acid batteries, but when used in forklifts, the lighter weight can be a disadvantage as weight is needed to provide stability.

- **Delivery**—While many of us have heard about (and maybe even seen) air drones in use for recreational, military, and now potentially product delivery by Amazon, the latest drones are actually land based.

  For example, there's a land-based drone from a startup called Starship that has already been tested in the UK and is soon to be tested in the United States. Another one from a startup called Marathon has been tested by Domino's in Australia for pizza delivery. Furthermore, whether the delivery comes via traditional means or via drone, companies such as Amazon are also allowing customers to trigger orders for local delivery using apps and, in Amazon's case, a voice-activated bar code scanner called Dash.

While these technologies might seem far-fetched or at least far out, there is no doubt that the "last mile" of retail will be radically transformed in the not-too-distant future.

These changes will also have a huge impact on the leanness of companies' supply chains. Using a land drone is expected to cost one-tenth as much as traditional (local) delivery methods. On top of that, we all know the impact that point-of-sale (POS) data in combination with collaborative programs has had on improving efficiencies in the supply chain in many industries. Can you imagine the added benefit of manufacturers and distributors knowing not only inventory levels in their customers' warehouses through current collaborative programs but potentially in consumers' homes as well? They will also have the added ability to capture actual demand at its source for further improvements in forecasting [Myerson, 2016].

Hardware Technologies for a Competitive Advantage in the Next Decade

According to a 2016 MHI annual industry report [Columbus, 2016], 83% of survey respondents said they believe that at least one of the following eight technologies (mostly hardware based or related) could either be a source of competitive advantage or a source of disruption for supply chains in the next 10 years:

- Predictive analytics
- Robotics and automation
- Sensors and automatic identification
- Wearable and mobile technology
- Driverless vehicles and drones
- Inventory and network optimization tools
- Cloud computing and storage
- 3D printing

According to the survey, robotics and automation (51%), inventory and network optimization tools (48%), sensors and automatic identification (47%), and predictive analytics (44%) will deliver the greatest competitive advantage or disruption in the next 10 years.

The study also found that driverless vehicles and drones (50%) and 3D printing (48%) have the highest predicted adoption rates over the next 10 years, while cloud computing and storage (45%) and sensors and automatic identification (44%) are also being adopted quickly [Columbus, 2016].

Future Supply Chain Technology Spending

While changes and improvements in technology are being made rapidly, as discussed above, resources are always limited in an organization. It is therefore useful to see where money is currently being spent and where organizations plan to spend it in the future.

A 2016 Peerless Research Group (PRG) survey [Bond, 2016] found that anticipated investments in physical equipment like totes, racks, lift trucks, packaging, cranes, motors, and belts have tapered somewhat, while information technology is expected to remain a priority as respondents pursue mobile and wireless capabilities (30%), fulfillment solutions (23%), systems solutions (20%), and RFID solutions (19%). It appears that restrained spending on hardware is balanced by steady investment in IT solutions that tie equipment together.

More than one-third of respondents expect to handle smaller, more frequent orders, and 26% anticipate a growing need for multi-channel fulfillment capabilities. This could indicate that the increased focus on IT solutions is aimed in part at managing and leveraging store inventory to improve service levels.

On the software end of technology spending, the same survey found that 60% of respondents were currently using a warehouse management system (WMS). Among the more than half who planned to buy new WMS software, objectives included improved inventory deployment (36%), real-time control (34%), and labor management functionalities (25%). Roughly 54% intended to upgrade existing solutions.

Survey respondents who were already using supply chain management software identified the most important initiatives for their current needs: demand planning (66%), inventory visibility (62%), procurement (60%), and collaborative forecasting, planning, and replenishment (52%) [Bond, 2016].

Looking Ahead

A research report from the Institute for Global Futures on the future of supply chains and global business [Canton, 2011] concluded that we are headed toward an era of greater complexity, more competition, and faster change. It makes the following predictions:

- Innovations will come sooner, and technology will continue to be a prime enabler of market share growth, profitability, and competitive advantage. Information technology (IT) will become even more important to the survival of organizations.
- Supply chains are becoming more proactive, predictive, and anticipatory, with a key trend of electronic collaboration as a technology to differentiate their value.
- Web-centric end-to-end solutions will become tomorrow's EDI, as smarter and faster tools and applications and networks emerge.

- Supply chain partners that invest in flexible, scalable, and adaptable tools will succeed, and real-time information will be a key differentiator that will eventually be a standard for all supply chains.

The research points to seven paradigm shifts that can be expected in future supply chains:

- **Real-time predictive forecasting**—Enterprises will be able to anticipate and forecast future customer demand with real-time data to identify profitable niche markets that they can quickly turn into products and services.

- **Business intelligence**—Information from a variety of sources, internal and external, about competitors, customers, and the industry will be used to quickly identify and develop profitable opportunities.

- **On-demand service**—Supply chain resources will be designed and implemented not based solely on cost or efficiency but also based on fast, smart software, and system optimization to enable access to a flexible, transparent, and interoperable knowledge network.

- **Pervasive networking**—Intelligent devices will be placed in packaging to monitor status via mobile devices. These devices will be pervasive, intelligent, and fully integrated into the products being moved in the supply chain to provide efficiency, speed, and knowledge.

- **Electronic markets**—Electronic markets will emerge, aided by artificial intelligence (AI), where price, speed, and feature transparency will dominate. These markets will present new opportunities as interactive web TV and wireless commerce become widespread in use throughout the world.

- **Smarter software**—Software will automate human functions with less error and more speed to enable more efficient and more cost-effective supply chains with many functions automated around rule-based AI systems, providing faster efficiencies.

- **Next-generation collaborative IT infrastructure**—In the future, there will be deeper collaboration, offering transaction, communications, clearing, confirmation, validation, and decision support with customers, suppliers, partners, and even competitors. Agility and the intelligence of collaboration will determine success. The delivery channel will be tied to customers, and data warehouses will be linked to the desktop or wireless web, with real-time, on-demand streaming data.

According to research conducted by the Institute for Global Futures, the following are some of the prime enablers that will shape the future of collaborative supply chains:

- Artificial intelligence decision support for predictive modeling and automatic logistics management

- Innovations such the next-generation Internet; broadband; peer-to-peer; nanotechnology; grid computing (a collection of computer resources from multiple locations to reach a common goal); the semantic web to promote common data formats and exchange protocols and VoIP to provide voice and data integration with wireless, satellite, and interactive TV for improved logistics coordination

- On-demand supply chain components that are created and then become invisible until needed

- Banking and telecom infrastructure tied invisibly to the enterprise and to logistics systems

- Just-in-time production that is tied to predictive analysis of customer demands and logistical systems availability

- Real-time data mining to drive the optimal utilization of supply chain resources, systems, and costs

- Customer supply chain hosting/sharing to repurpose underutilized infrastructure within and outside the customer's systems

- Supply chain optimization objects built into the system with digitally engineered personalities (DEPS) that automatically monitor and configure customer, logistic, and supplier needs

- What-if scenarios with forward-placed digital cash commitments to determine buyers' interest before manufacturing starts

In summary, this research predicts that the era of the supply chain that reacts to customer needs alone is over, and leaders are the supply chain participants who understand this paradigm shift [Canton, 2011].

While many of these predictions about the future of the supply chain and its associated technology may morph over time, and some not come to fruition at all, there is no doubt that we live in exciting, fast-changing, volatile times, and organizations that use technology to enable an agile, flexible, Lean supply chain will be the winners of tomorrow.

References

Chapter 1

"Capital Expenditures on Technology Driving 2016 Financial Plans, CFOs Report in TD Bank Survey," TD Bank News Release, www.tdbank.com, November 16, 2015.

Gartner, "Gartner Says Worldwide IT Spending on Pace to Grow 2.4 Percent in 2015," Press Release, January 12, 2015. Last accessed at www.gartner.com.

Aberdeen Group, "Global Supply Chain Benchmark Report," June 2006. Last accessed at www-935.ibm.com, 2015.

Heaney, Bob, "Supply Chain Visibility," Aberdeen Group, 2013. Last accessed at www.gs1.org, 2016.

Krajewski, Lee J., Larry P. Ritzman, and Manoj K. Malhotra, *Operations Management: Processes and Supply Chains*, 10th ed., Pearson, 2013.

Krigsman, Michael, "Compelling Advice for the CFO," February 22, 2013. Last accessed at www.zdnet.com.

Ariba, "Mitigate Supply Chain Risk with Collaboration and Visibility to Achieve the Perfect Order," November 3, 2014. Last accessed at www.ariba.com.

Pisello, Thomas, *Return on Investment—For Information Technology Providers*, Information Economics Press, 2001, p. 1.

Porter, Michael, *Competitive Advantage: Creating and Sustaining Superior Performance*, The Free Press, 1985.

Porter, Michael, *Competitive Strategy: Techniques for Analyzing Industries and Competitors*, The Free Press, 1998.

Ramakrishman, Sreekanth, and Michael Testani, "People, Process, Technology—The Three Elements for a Successful Organizational Transformation," IBM SEMS Webinar, March 2, 2011.

"SCOR Model," www.supply-chain.org. Last accessed 2015.

Chapter 2

Bozarth, Cecil, and Robert Handfield, *Introduction to Operations and Supply Chain Management*, 2nd ed., Pearson, 2008.

Capgemini Consulting, "How Will Digital Impact SCM: Supply Chain Trends," September 9, 2014. Last accessed at www.capgemini.com, 2015.

Council of Supply Chain Management Professionals (CSCMP), "CSCMP Supply Chain Management Definitions and Glossary." Last accessed at www.cscmp.org, 2015.

Fawcett, S. E., and G. M. Magnan, "The Rhetoric and Reality of Supply Chain Integration," *International Journal of Physical Distribution & Logistics Management*, Vol. 32, No. 5, 2002, 339–361.

Simatupang, Togar M., and R. Sridharan, "A Characterization of Information Sharing in Supply Chains," Massey University, October 2001. Last accessed at www.academia.edu, 2015.

Trunick, Perry A., "Continuing Education—Making the Right Selection," *Inbound Logistics*, February 2011. Last accessed at www.inboundlogistics.com, 2011.

U.S. Chamber of Commerce, "Global Supply Chain, Customs and Trade Facilitation." Last accessed at https://www.uschamber.com/issue-brief/global-supply-chain-customs-and-trade-facilitation.

Chapter 3

Butcher, David, "Technology's Role in Lean Today," Thomasnet Industry News, November 10, 2009. Last accessed at www.thomasnet.com, 2015.

Martichenko, Robert, "The Lean Supply Chain: A Field of Opportunity," *Inbound Logistics*, January 2013. Last accessed at www.inboundlogistics.com, 2015.

Myerson, Paul, "How to Cut Seven Non-Traditional Wastes," *Inbound Logistics*, June 2015.

"How Much Technology Is Needed for Lean Manufacturing?" *Supply Chain Digest*, March 9, 2011. Last accessed at www.scdigest.com, 2015.

Thompson, Richard, Karl Mankrodt, and Kate Vitasek, "Lean Practices in the Supply Chain," Jones Lang LaSalle, 2008. Last accessed at www.joneslanglasalle.com, 2015.

Trent, Robert J., *End-to-End Lean Management: A Guide to Complete Supply Chain Improvement*. J. Ross Publishing, 2008.

Womack, James P., and Daniel T. Jones, *Lean Thinking: Banish Waste and Create Wealth in Your Corporation*, 2nd ed., Productivity Press, June 1, 2003.

Chapter 4

Bozarth, Cecil, and Robert Handfield, *Introduction to Operations and Supply Chain Management*, 2nd ed., Pearson, 2008.

Feldman, Jonathan, "Research: 2012 Enterprise Project Management," *InformationWeek*, January 2012.

Gartner, "Gartner Says Worldwide Supply Chain Management Software Market Grew 7.1 Percent to Reach $8.3 Billion in 2012" (Press Release). Last accessed at www.gartner.com, 2014.

Gilmore, Dan, "Insight from the 2010 Gartner Supply Chain Study," *Supply Chain Digest*, June 8, 2010. Last accessed at www.scdigest.com, 2014.

Gilmore, Dan, "Insight from the 2013 Gartner Supply Chain Study," *Supply Chain Digest*, June 28, 2013. Last accessed at www.scdigest.com, 2014.

Harrington, L. (2007). "Defining Technology Trends," *Inbound Logistics*, April 2007.

Heizer, Jay, and Barry Render, *Operations Management*, 11th ed., Pearson, 2013.

McDonnell, R., E. Sweeney, and J. Kenny, "The Role of Information Technology in the Supply Chain," *Logistics Solutions*, Vol. 7, No. 1, 2004, 13–16.

Simatupang, Togar M., and R. Sridharan, "A Characterization of Information Sharing in Supply Chains," Massey University, October 2001. Last accessed at www.academia.edu, 2015.

erpsearch, "Supply Chain Management Software White Paper." Last accessed at www.erpsearch.com, 2014.

Chapter 5

Po, Vincent, "Understanding the 3 Levels of Supply Chain Management," *The Procurement Bulletin*, December 12, 2012. Last accessed at www.procurmentbulletin.com, 2015.

Spinnaker, "Introduction to Strategic Supply Chain Network Design, Perspectives and Methodologies to Tackle the Most Challenging Supply Chain Network Dilemmas" (White Paper and Case). Last accessed at www.spinnakermgmt.com, 2015.

Establish Inc., "Client Case Study—Global Network Design" (White Paper). Last accessed at www.establishinc.com, 2015.

Chapter 6

Logility, "Caribou Coffee Case Study." Last accessed at www.logility.com, 2015.

Harris, Daniel, "Compare Demand Planning & Forecasting Software," November 15, 2015. Last accessed at www.softwareadvice.com, 2015.

Kahn, Kenneth B., and John Mello, "Lean Forecasting Begins with Lean Thinking—On the Demand Forecasting Process," *Journal of Business Forecasting*, Winter 2004–2005, pp. 30–32, 40.

Cooke, James A., "Kimberly-Clark Connects Its Supply Chain to the Store Shelf," *Supply Chain Quarterly*, Quarter 1, 2013. Last accessed at www.supplychainquarterly.com, 2015.

JDA Software, "Recipe for Success" (Case Study). Last accessed at www.jda.com, 2015.

SAS, "The Lean Approach to Business Forecasting—Eliminating Waste and Inefficiency from the Forecasting Process" (White Paper), 2012. Last accessed at www.sas.com, 2015.

Chapter 7

Bartholomew, Doug, "Can Lean and ERP Work Together?" *IndustryWeek*, April 12, 2012.

Corporate Technologies, "Case Study: Manufacturing Production Planning." Last accessed at www.cptech.com, 2015.

ORM Technologies, "Production Planning Case Study." Last accessed at www.orm-tech.com, 2015.

Salman, Mustafa Ramzi, Roman van der Krogt, James Little, and John Geraghty, "Applying Lean Principles to Production Scheduling," 2010. Last accessed at www.researchgate.net.

Techlogix, "Techlogix Helps Nestlé Innovate in Milk Production Planning" (Case Study). Last accessed at www.techlogix.com, 2015.

Chapter 8

Continental Mills, "Logility Voyager Solutions Case Study." Last accessed at www.logility.com, 2015.

Dougherty, John, and Christopher Gray, *Sales and Operations Planning—Best Practices*. Trafford Publishing, 2006.

JDA, "Infinite Possibilities: Infineon Technologies Takes Planning to the Next Level with JDA S&OP" (Case Study), 2014. Last accessed at www.jda.com, 2015.

Lance, "Logility Voyager Solutions Case Study." Last accessed at www.logility.com, 2015.

Viswanthan, Nari, "S&OP—Strategies for Managing Complexities with Global Supply Chains," Aberdeen Group, 2010. Last accessed at www.aberdeen.com, 2015.

"Leading High Tech Company Has to Improve S&OP Process, Supporting Tools Rapidly After Outsourcing Strategy Leads to Real Challenges," *Supply Chain Digest*, July 24, 2013. Last accessed at www.scdigest.com, 2015.

Chapter 9

"A Picture Perfect MRP Implementation Helps Traffic Enforcement Camera Maker to Profitability", Case Studies, www.ez-mrp.com, 2016.

Abilla, Pete, "How Manufacturing Software Can Adjust to Lean Principles," 2012. Last accessed at www.schmula.com, 2016.

Benton, W.C., and Hojung Shin, "Manufacturing Planning and Control: The Evolution of MRP and JIT Integration," *European Journal of Operational Research*, Vol. 110, No. 3, November 1998, pp. 411–440.

Gables Engineering, "Case Study, IFS Software." Last accessed at www.top10erp.org, 2016.

Kowalke, Mae, "5 Trends in MRP Technology," *Inside-ERP*, July 21, 2015. Last accessed at www.it.toolbox.com, 2015.

Ptak, Carol, and Chad Smith, "Lean Finds a Friend in Demand Driven MRP (DDMRP)." Last accessed at www.demanddriveninstitute.com, 2012.

Exostar Raytheon, "Raytheon Streamlines and Automates Its Material Requirement Planning Processes with Exostar's Supply Chain Platform." Last accessed at www.exostar.com, 2016.

Chapter 10

SAP, "Clariant Cuts Costs with Ariba Solutions—Automating and Enhancing Procurement Processes." Last accessed at www.sap.com, 2016.

Dominick, Charles, "Ten Types of Procurement Software," July 1, 2015. Last accessed at www.webcom.com, 2016.

JDA, "Enabling Online Supplier Collaboration at Toshiba Semiconductor Company." Last accessed at www.jda.com, 2016.

Heizer, Jay, and Barry Render, *Operations Management*, 10th ed., Pearson, 2011.

Basware, "New Purchase-to-Pay System Allows Smarter Processes at Atea." Last accessed at www.basware.com, 2016.

Chapter 11

IFS Software, "10 Ways to Use ERP to Lean the Manufacturing Supply Chain" (White Paper), 2009. Last accessed at www.ifsworld.com, 2016.

Ultra Consultants, "Case Study: Radio Flyer," 2008. Last accessed at www.ultraconsultants.com, 2016.

IFS ERP Systems, "Case Study: Flexpipe Systems Inc.," 2015. Last accessed at www.top10erp.org, 2016.

IQMS Manufacturing, "Nissen Chemitec America—Leading Automotive Supplier Accelerates Lean Operations with IQMS ERP" (Case Study), 2015. Last accessed at www.iqms.com, 2016.

Schiff, Jennifer Lonoff, "9 Tips for Selecting and Implementing an ERP System," *CIO*, July 30, 2014. Last accessed at www.cio.com, 2016.

Chapter 12

Rockwell Automation, "Full Sail Brewing Taps Manufacturing Intelligence to Enhance Brewing Process" (White Paper), September 2011. Last accessed at www.rockwellautomation.com, 2016.

CI Precision, "CI Precision Implements First Suite of Ci-DMS Package in Asia/Pacific Region" (Case Study). Last accessed at www.ciprecision.com, 2016.

Cottyn, Johannes, Hendrik Van Landeghem, Kurt Stockman, and Stijn Derammelaere, "The Role of Change Management in a Manufacturing Execution System," *Proceedings of the 41st International Conference on Computers & Industrial Engineering.* Last accessed at www.usc.edu, 2016.

EazyWorks, "EZ-MES Production Tracking System: Case Study." Last accessed at www. eazyworks.com, 2016.

Chapter 13

"Auto Parts Manufacturer Chooses Asprova for Its Good User Interface Reduces Labor of Adjusting the Schedule" (Case Study). Last accessed at www.asprova.com, 2016.

Kreipl, Stephan, and Michael Pinedo, "Planning and Scheduling in Supply Chains: An Overview of Issues in Practice," *Production and Operations Management Society (POMS)*, Vol. 13, No. 1, Spring 2004, pp. 77–92.

"Mueller Stoves Reduces the Assembly Line Stops After Preactor Deployment" (Case Study). Last accessed at www.preactor.com, 2016.

Chapter 14

Banker, Steve, "Return on Investment for Transportation Management Systems," *ARC Strategies*, November 2011. Last accessed at www.leanlogistics.com, 2016.

"Everlast Builds a Championship Company with New Product Lines," *DS Magazine*, Vol. 7, No. 1, Spring 2007, pp. 3–5.

Turbide, David, "How Can Distribution Requirements Planning Help Inventory Management?" Last accessed at www.searchmanufacturingerp.techtarget.com, 2016.

Martichenko, Robert, "Lean Transportation Management: Creating Operational and Financial Stability." Last accessed at www.leancor.com, 2016.

Murphy, Jean V., "Canadian Tire Keeps Stores Rolling with Replenishment Program," *Supply Chain Brain*, October 1, 1999. Last accessed at www.supplychainbrain.com, 2016.

Partridge, Amy Roach, "Auto Logistics: Revving Up Service Parts Logistics Operations," *Inbound Logistics*, January 2011.

Chapter 15

Ultra Ship TMS, "Leading Dairy Trims 18% from Transportation Costs Using Optimizer Software" (Case Study), 2015. Last accessed at www.ultrashiptms.com, 2016.

Transwide TMS, "Miller Brands UK Uses Transwide TMS to Manage Growing Transport Volumes" (Case Study). Last accessed at www.transwide.com, 2016.

Manhattan Associates, Inc., "Papa on the Platform—Hold the Anchovies: Papa John's Pizza Orders Optimization Supreme With Manhattan's Supply Chain Process Platform" (Case Study), 2013. Last accessed at www.manh.com, 2016.

Chapter 16

Meller, Russ, "Order Fulfillment as a Competitive Advantage," *Supply Chain 24/7*, March 5, 2015. Last accessed at www.supplychain247.com, 2016.

Cadre Technologies, "TAGG Logistics—Supply Chain Management & Order Fulfillment" (Case Study). Last accessed at www.cadretech.com, 2016.

Manhattan Associates, Inc., "Whirlpool Spins Optimized Supply Chain with Help from Manhattan Associates" (Case Study), 2013. Last accessed at www.manh.com, 2016.

Chapter 17

Roger, Dale S., and Ronald S. Tibben-Lembke, *Going Backwards: Reverse Logistics Trends and Practices*, Reverse Logistics Executive Council, 1998.

RTL, "RTL™ Adds Up to Many Happy Returns for Philips" (Case Study). Last accessed at www.roundtriplogistics.com, 2016.

Return Pool, "Nokia—Deployment of Global Service Strategy" (Case Study), 2012. Last accessed at www.returnpool.com, 2016.

GENCO, "Return Central—Reconstructed Network Yields Big Efficiencies, Faster Processes" (Case Study). Last accessed at www.genco.com, 2016.

Chapter 18

Bain & Company, "Survey of 300 Global Companies." Last accessed at www.bain.com, 2016.

W/SAS, "Improving Performance Through Predictive, Data-Driven Insights Supply-Chain Analytics: Beyond ERP & SCM," 2010. Last accessed at www.sas.com/supplychain.

Noyes, Katherine, "Why Analytics Is Eating the Supply Chain," *Computerworld*, April 29, 2016.

Partridge, Amy Roach, "Business Intelligence in the Supply Chain," *Inbound Logistics*, April 2013. Last accessed at www.inboundlogistics.com, 2016.

Chapter 19

Kinaxis Customer Spotlight, "Agilent Technologies: Comprehensive Supply Chain Visibility Across a Multi-Enterprise Supply Chain," 2014. Last accessed at www.kinaxis.com, 2016.

One Network Enterprises, "Arrow Electronics Automates Replenishment Program with One Network's Real Time Value Network" (Case Study), 2011. Last accessed at www.onenetwork.com, 2016.

Myerson, Paul, "Supply Chain Integration + Collaboration = Time Travel?" *Inbound Logistics*, December 2014.

Chapter 20

PNC Bank, "4 Trends in Supply Chain Software and Systems." Last accessed at www.pnc.com, 2016.

Abrosimova, Kate, "How You Can Develop Uber for Trucking," December 14, 2014. Last accessed at www.medium.com, 2016.

Bond, Josh, "2016 Warehouse/DC Equipment Survey: Investing in Information Infrastructure," *Modern Materials Handling Magazine*, April 1, 2016.

Canton, James, "The Future of Collaborative Supply Chains and Global Business," 2011. Last accessed at www.globalfuturist.com, 2016.

Columbus, Louis, "Eight Technologies Revolutionizing Supply Chains," *Forbes*, April 10, 2016. Last accessed at www.forbes.com, 2016.

Felgendreher, Boris, "The 11 Supply Chain Trends Businesses Cannot Ignore," February 2014. Last accessed at www.cips.org, 2016.

Heistand, Steve, "Five Trends in Supply Chain Management (SCM) Software." Last accessed at www.erpsearch.com, 2016.

Myerson, Paul, "Planes, Trains, Automobiles[el]and Drones?" *Industry Week*, April 6, 2016.

Index

REGISTER YOUR PRODUCT at informit.com/register
Access Additional Benefits and SAVE 35% on Your Next Purchase

- Download available product updates.

- Access bonus material when applicable.

- Receive exclusive offers on new editions and related products.
  (Just check the box to hear from us when setting up your account.)

- Get a coupon for 35% for your next purchase, valid for 30 days. Your code will
  be available in your InformIT cart. (You will also find it in the Manage Codes
  section of your account page.)

Registration benefits vary by product. Benefits will be listed on your account page
under Registered Products.

InformIT.com–The Trusted Technology Learning Source
InformIT is the online home of information technology brands at Pearson, the world's foremost
education company. At InformIT.com you can

- Shop our books, eBooks, software, and video training.
- Take advantage of our special offers and promotions (informit.com/promotions).
- Sign up for special offers and content newsletters (informit.com/newsletters).
- Read free articles and blogs by information technology experts.
- Access thousands of free chapters and video lessons.

Connect with InformIT–Visit informit.com/community
Learn about InformIT community events and programs.